Ancient Chinese Bronzes

in The Saint Louis Art Museum

ANCIENT CHINESE BRONZES

IN THE SAINT LOUIS ART MUSEUM

Steven D. Owyoung

Technical Observations and Analyses by
Suzanne Hargrove

Essay by
Thomas Lawton

This publication was funded, in part, by the Andrew W. Mellon
Foundation. Additional financial assistance was provided by the
National Endowment for the Arts, a federal agency.

Managing Editor: Mary Ann Steiner
Editors: Kathleen M. Friello and Reiko Tomii
Designer: Andy Palombo

Printed and bound by Scholin Brothers Printing Company in the U.S.A.

First edition

Published in the United States by The Saint Louis Art Museum,
#1 Fine Arts Drive, St. Louis, Missouri 63110-1380.

Library of Congress Cataloguing-in-Publication Data

Owyoung, Steven D.
    Ancient Chinese bronzes in The Saint Louis Art Museum / texts by
  Steven D. Owyoung, Suzanne Hargrove, and Thomas Lawton.
      p. cm.
    Includes bibliographical references and index.

ISBN 0-89178-045-9
Library of Congress Catalog Card Number: 97-066604

Cover: Detail of *Gu*, 12th century BC, Shang dynasty, height: 32.7 cm,
Gift of J. Lionberger Davis 217:1950 (No. 8)

CONTENTS

8   Foreword
     James D. Burke

9   Acknowledgments

11   Introduction
     Steven D. Owyoung

14   Notes to the Reader

17   Chinese Ritual Bronzes: Collections and Catalogues Outside China
     Thomas Lawton

38   Catalogue of Bronzes
     Steven D. Owyoung

148 Technical Observations and Analyses of Chinese Bronzes
     Suzanne Hargrove

184 Index

Foreword

A museum holds in public trust many precious elements of the past; works of known value, and works which mature in our appreciation and continue to evolve under every new inspection. The collection of ancient Chinese bronzes held by The Saint Louis Art Museum has steadily grown, in both size and significance, since its inception in 1916.

For the Museum, this is the first publication of the bronzes as a group since J. Edward Kidder, Jr.'s *Early Chinese Bronzes* in 1956. We welcome the opportunity to present this portion of the collection and a new consideration of these works of art. The present volume contains a valuable range of information, placing the bronzes in a variety of contexts. Each section contributes to a fuller knowledge of the works, their original identity, their physical history as surviving artifacts of great age, and their value as objects of appreciation and acquisition.

The major part of this catalogue was written by our Curator of Asian Art, Steven D. Owyoung. It offers a detailed description of each object's appearance, its salient features of style, form, and decoration, and its relationship to other known examples. The revelations and mysteries of the inscriptions borne by many of the vessels are also examined. A history of connoisseurship and catalogues in Japan, Europe, and the United States charting the evolution of Chinese bronze collecting and knowledge in the West was prepared by Thomas Lawton, Senior Research Scholar of the Freer Gallery of Art. Suzanne Hargrove, the Museum's Conservator of Objects, has provided technical analyses of selected works that offer a different history of the objects: evidence of their physical life, original composition, and the effects of time, situation, and the hands of other conservators and collectors over the ages.

This publication was conceived over ten years ago and has benefitted from the careful thought and research of many individuals. Our particular thanks are expressed in the acknowledgments which follow. The catalogue was generously funded by the Andrew W. Mellon Foundation and the National Endowment for the Arts.

James D. Burke
Director
The Saint Louis Art Museum

ACKNOWLEDGMENTS

In my study of Chinese art I owe much to many people: but here I would like to thank a few of those who contributed to my appreciation for ancient bronzes. I first learned of bronzes from my early teachers, Chiang Yee of Columbia University and Hon-Chew Hee, my granduncle, who taught me the epigraphic and calligraphic importance of bronze inscriptions. When I arrived in Taipei, Marshall Wu, then Chief of Exhibitions, and Na Chih-liang, then Curator of Calligraphy and Painting and my teacher in archaic jades, introduced me to the bronzes in the imperial collection at the National Palace Museum in Taiwan. When I joined the Palace Museum staff, the bronze specialists T'an Tan-chiung, Curator of Antiquities, and Vice-Director Ho Lien-kwei were among my tutors in connoisseurship and art history. I often visited Yuan Chai, Head Conservator, and the other conservators and artist-technicians who were casting bronzes to learn more about the physical and technical properties of the ancient medium.

Among Taipei's permanent displays were many materials from the archaeological excavations at Anyang conducted by the Academia Sinica from 1928 through 1937. The significance of the Anyang finds and later archaeology was reinforced by my studies with Professor Richard Edwards and Professor Virginia Kane at the University of Michigan. Kane impressed on me the importance not only of precision and analysis, but also intellectual exploration. During my graduate fellowships in Chinese painting at the Freer Gallery of Art, I spent my spare time in the galleries and storerooms looking at jade and bronze works. On these occasions, which could have been considered truancy from my painting studies, Thomas Lawton, then Director of the Freer, would appear and say a provocative word or two about a piece before leaving me to puzzle out the rest. It was Lawton who first encouraged me to write a catalogue of the Museum bronzes when I arrived in St. Louis in 1983. In the Freer's laboratories, I was often lucky enough to find W. Thomas Chase, Head Conservator, willing to share his latest test results.

The late Max Loehr, renowned bronze specialist and Professor of Chinese art at Harvard, had already retired when I joined the Fogg Art Museum in Cambridge, but he occasionally dropped by to "chat" whenever I exhibited the Winthrop bronzes and archaic jades. Even during these casual conversations, Loehr's descriptions of artworks were models of exactitude and illumination. Under John Rosenfield, then Acting Director of the Fogg, and Abby Aldrich Rockefeller, Professor of Oriental Art, Dawn Ho Delbanco and I worked together on *Art from Ritual*, an exhibition and symposium on the Arthur M. Sackler collection of ancient Chinese bronzes.

While a painting specialist by training, I am partial to the bronze medium, in both its ancient and later manifestations, and willingly succumb to the ample seductions of the art form. This study, which has furnished me the opportunity to research and write about an especially fine group of works, is offered to all those who are interested in the character and inimitable beauty of Saint Louis's ancient Chinese bronzes. Both Lawton and Chase have kindly read the entries and made valuable suggestions. I have had the benefit of their advice, but errors great and small are mine alone.

I am grateful to Thomas Lawton and Suzanne Hargrove, whose additional texts greatly expanded the scope of this book. Lawton, former Director and Curator of the Freer Gallery of Art, contributed a history of Chinese bronze collecting in the West. He provides the field with a detailed and penetrating look at the many scholars, collectors, and institutions responsible for our appreciation of ancient bronzes today. Hargrove, Head Conservator of Objects at the Museum, enthusiastically participated in the technical side of the project and has contributed her knowledge and energies to this study. Her analyses of selected bronzes from the collection will afford scholars and conservators a more profound understanding of the physical qualities of these works, both visible and invisible.

The project benefited from the visit in 1994 by W. Thomas Chase, Head Conservator at the Freer Gallery of Art and Arthur M. Sackler Gallery, and Zhou Bao Zhong, Director of Department of Conservation, National Museum of Chinese History, Beijing, who spent about a week examining the Saint Louis bronzes. The technical information in this book is based in part on their commentary. Chase was especially helpful in interpreting X-radiographs and lead isotope data on some of our most interesting bronzes.

Over the years, Director James D. Burke and the Museum staff have provided assistance and encouragement on this project, one of many programs that reveals the extraordinary depth of our collections. Diane Burke, Museum Conservator, provided me with the first close look at the Museum's bronzes during conservation for their reinstallation in 1987. The remarkable condition of the vessels is due to her formidable skills and judicious eye.

Research for this book has been conducted with the assistance of several individuals. During their studies at Washington University, Katherine Wendel and Lisa L. Somogyi did research in comparative materials as related to the Museum bronzes early in the project. Wendel and Somogyi were especially interested in the bronzes related to archaism during the Shang period. I am most grateful to Melissa Bowen-Li, a superb researcher and manager who gave structure to literally all of the information and files for the work. Thanks go to Christa Kirby, Museum Intern from Washington University, who assisted with library and reference searches.

Much appreciation and many thanks go to Pat Woods, Photographic Manager, who handled the photographic project with her typical efficiency and engaging humor. She and Donald Straussner, Museum Installer, learned more about graph-vessel alignment than they ever wanted to know and kept photographer Bob Kolbrener busy shooting bronze after bronze. Kolbrener, with his probing eye and magician's camera, was able to capture not only the aesthetic character of the ritual bronzes but also the data and detail of every object.

My special thanks to Mary Ann Steiner, Director of Publications, for her faith and tenacity in steering the project through to completion. Steiner was ably assisted by Suzanne Tausz and Jill Henderson who shepherded the manuscripts to and fro the widening electronic maze. I am especially thankful to Kathleen M. Friello and Reiko Tomii who edited the whole manuscript with great professionalism and sharp scholarly attention; they also created the index, which will serve as a valuable tool for all who use this book. Andy Palombo contributed the handsome design that enhances the beauty and splendor of the Museum's Chinese bronzes.

S. O.

INTRODUCTION

Steven D. Owyoung

The Chinese bronze collection of The Saint Louis Art Museum is a strong and important part of its distinguished holdings in Asian art. For more than eighty years, the Museum has actively acquired bronzes and the other arts of China, including sculpture, ceramics, calligraphy, and painting. In actual number, the Museum's bronze holdings are relatively modest, but, as with other parts of the Asian collection, the importance of individual works is noteworthy, and there are several bronzes of exceptionally high quality.

The development of the bronze collection has been influenced by history, events, and people. Its continued growth and potential are greatly due to the interest of the Museum's patrons, collectors, and connoisseurs, both past and present. At the turn of the century, active American interest in Chinese art was strong in New York, Boston, Detroit, and Chicago. Although it is likely that there was a growing appreciation of Asian art in Saint Louis during the late 19th century, it is certain that sustained interest in Chinese art and bronzes in particular was forged with the advent of the St. Louis World's Fair.

The Louisiana Purchase Exposition of 1904, which was held in Forest Park, was the setting for China's official inaugural entry to cultural events organized on such a global scale. Prince Pulun (1875–?), a member of the royal family, was sent as Imperial High Commissioner to the Fair. China's president ex officio of its exhibits was no less a historic figure than Sir Robert Hart (1835–1911), Inspector-General of the Chinese Imperial Maritime Customs Service and Ambassador of Great Britain to China. The emperor, Puyi (1906–67), sent as commemorative gifts a jade scepter and a carved boulder of river jade. The Chinese government appropriated an amount equal to one half million dollars for its participation. Among the highlights of the Chinese exhibits was a reproduction of Prince Pulun's summer residence in Beijing, a complex of three fully furnished main buildings, which the Prince presented to the president of the Exposition at the close of the Fair. China had only a very small number of art exhibits in the Palace of Fine Arts, but its extensive displays in the Palace of Liberal Arts included many relics from its ancient past. Archaic jades, rare porcelains, and ancient bronzes were focal points of many of the Chinese exhibits. Ritual bronzes were shown by a number of state, provincial, municipal, and commercial sources, including the Industrial Institute of Beijing, the province of Hubei, the city of Canton, Wing Cheong and Company, and other exhibitors.

An extraordinary entry among the groups of Chinese ancient art shown at the Exposition was the distinguished private collection of the high government official Duanfang (1861–1911). Renowned as a fine collection of paintings, porcelain, and bronze, Duanfang's holdings epitomized classical Chinese taste and connoisseurship. The Viceroy's bronzes included important works, such as the famous Baoji altar set and the Saint Louis *jia* (No. 20; 221:1950), but it is still unclear as to whether these were exhibited at the Fair. His collection was part of the displays in the Palace of Liberal Arts for Hubei, the Chinese province of which Duanfang was governor at the time.

The presence of the Duanfang collection at the Fair undoubtedly excited the interest of those knowledgeable in Chinese art, the foremost among them being the St. Louisan William K. Bixby (1857–1931) and Charles Lang Freer (1854–1919) of Detroit. Bixby was president of the American Car and Foundry Company, a manufacturer of railroad cars. He was also an active patron of many of St. Louis's cultural institutions and an energetic force behind the Asian collections at the Museum in its early years. Freer, also a railroad man, and a collector of Chinese art from 1893, gave his collection of Asian and

American art to the Smithsonian where it is housed and on view at the Freer Gallery of Art. The two men were closely associated with the Exposition: Bixby was Director of the Fair and Chairman of the Committee on Fine Arts; Freer was a member of the National Advisory Committee for the Fair's Department of Art. Bixby and Freer, sharing a mutual interest in collecting and railroads, forged a strong friendship.

Within a year of his retirement from business in 1906, Bixby became President of the Museum Board. Under his long tenure the Museum acquired its first early Chinese bronzes, among which is the early Western Zhou *gui* (No. 26; 203:1916). In 1919, Bixby personally undertook the development of the Museum's Asian collections by taking a trip to Korea, Japan, and China to buy works of art. These purchases and later gifts to the Museum laid the foundation for the Saint Louis Asian collections.

The Chinese art dealer C.T. Loo (1880–1957), who had galleries in Paris and New York, was a frequent visitor to St. Louis from as early as 1920. One of the most important dealers in Chinese art in America at the time, Loo educated collectors to appreciate Shang and Zhou bronzes as well as Chinese ceramics and archaic jades by mounting museum and gallery exhibitions. After one such showing in 1940, the Museum purchased the early Western Zhou *fanglei* (No. 24; 2:1941).

The impact of C.T. Loo on collecting in St. Louis was strong and his influence is seen in the activities of J. William Beckmann (d.1954) and J. Lionberger Davis (1878–1973). Beckmann's sister, Leona J. Beckmann, devoted more than twenty-five years' volunteer service to the Museum; she, as well as her sister Anna, Mrs. Charles Kunkel, presented the Museum with bronzes from their brother's estate. J. Lionberger Davis, a longtime benefactor and advocate of the Museum, was a St. Louis lawyer who had been educated at Princeton and Harvard and graduated in 1903 from St. Louis Law School. Davis's involvement with the Museum as a collector of Chinese and Western art was passionate and long-lived. With a scholarly appreciation for bronzes and a connoisseur's eye, Davis acquired an impressive group of ancient Chinese bronzes of which he gave more than thirty pieces to the Museum between 1950 and 1965, including the rare coiled serpent (No. 30; 124:1951).

As the Museum's Chinese bronze collection grew through purchases and gifts, curator Thomas Temple Hoopes (1898–1981), a Harvard-trained specialist in arms and armor, saw to its care. Hoopes, who was a scholar of great breadth and interests, did most of the research and initial publication of the bronze acquisitions. Assisted by Catherine Filsinger Hoopes, whose faultless notes remain curatorial gems of object description, Hoopes provided the basis for the 1956 illustrated catalogue *Early Chinese Bronzes in the City Art Museum of St. Louis*, by J. Edward Kidder, Jr. of Washington University. With their publication, the Museum's bronzes drew the attention of the eminent bronze specialist Max Loehr (1903–88) of Harvard University, who published and included five of the Saint Louis pieces in *Ritual Vessels of Bronze Age China* at the Asia Society Gallery, New York in 1968, one of the most influential exhibitions of Chinese bronzes ever mounted.

During the 1970s and 1980s, there was a succession of important archaeological discoveries and exhibitions in the United States on ancient Chinese art from mainland China. These exhibits and the high quality of bronzes on the market, as well as the ongoing excavations in China, continue to influence the Museum's bronze collection and its development.

<table>
<tr><td>NOTES ON THIS BOOK</td><td>The forty-one ancient Chinese bronzes in this catalogue represent a major share of the important bronzes in The Saint Louis Art Museum's collection. Primarily a selection of the Museum's most notable ritual vessels, this group also consists of a small number of objects of functional, decorative, or symbolic significance. The majority of the bronzes are wine or water vessels; food containers and other works are in the minority but signifi-</td></tr>
</table>

cant. Fully half of the artworks date from the later Shang dynasty, mainly from the Anyang period. Reference to older motifs and designs is a particularly strong feature of the Museum's Shang bronzes, either as a continuance of developing styles or as archaism. Western Zhou bronzes, with a good representation of the transitional or early phase, dominate the remainder of the works.

From the beginning, the entries were written with the knowledge that fine color images would inform the reader as well as convey the special beauty of the bronzes. Detailed descriptions are given whenever the vessel type, shape, technique, motif, or inscription are particularly important to style, date, or interpretation. Because of their aesthetic importance, patina and corrosion have been described so that color, texture, and surface are more fully represented and appreciated. Inscriptions have been given for every inscribed vessel; the more important ones have been treated extensively. Each ritual vessel and artwork has been considered as an individual object, given special focus according to its most prominent element whether it be motif, design, shape, inscription, or vessel type. Comparative works, developments in vessel type or style, or historical discussions were brought to bear only when such issues furthered an understanding of the Museum's bronze.

A group of bronzes were selected for technical analysis. Each of these is described with consideration given to the original properties of the work; the alterations wrought on the bronze by time, physical circumstances, and various human agents; and its current discernable condition. A description of the bronze-casting process and a glossary of technical terms are also provided.

Numerous sources on bronzes written in Western languages, Chinese, and Japanese have been cited when they shed special light on a particular question. A major resource on Chinese bronzes is the three-volume catalogue of the Arthur M. Sackler collections written by Robert Bagley, Department of Art and Archaeology at Princeton University; Jessica Rawson, Warden of Merton College at Oxford University; and Jenny So, Curator of Ancient Chinese Art at the Freer Gallery of Art and the Arthur M. Sackler Gallery in Washington, D.C. Their work comprises the most comprehensive and in-depth scholarship on bronzes available in English and includes an enormous number of illustrated bronzes from collections around the world. The Sackler collections and catalogue provide a high standard for bronze research and formed the touchstone for this project.

Loehr's Typology

In a seminal study in 1953, Max Loehr proposed a scheme to describe the stylistic development of bronze surface design and the general decorative types of ancient Chinese bronzes. In a sequence that has weathered more than forty years of archaeological discoveries and, with minor modifications, is still widely accepted by the Chinese art field in America, Shang dynasty bronze decoration is arranged into five distinct styles.

Loehr Style I	Thin lines in thread relief on a plain ground representing simple forms in single bands in a decorative order that corresponds to early Erligang bronzes of the early middle Shang.
Loehr Style II	Wider relief ribbons representing thick, fluid forms with vertical divisions in a single band in a decorative order that corresponds to Erligang and Panlongcheng bronzes of the middle Shang.
Loehr Style III	Dense, relief lines representing highly ordered complex curvilinear forms with vertical divisions in compacted, multiple registers in a decorative order that corresponds to late Erligang and early Anyang bronzes of the late Shang.
Loehr Style IV	Broad ribbons with intaglio designs representing large forms or motifs clearly represented and distinguished from a ground of fine fretwork or spirals (*leiwen* or "thunder pattern") in multiple bands divided by vertical ridges or flanges in a decorative order that corresponds to late Erligang and early Anyang bronzes of the late Shang.
Loehr Style V	Motifs rendered in relief against a *leiwen* or smooth ground in single or multiple registers divided by ridges or flanges in a decorative order that corresponds to Anyang bronzes of the late Shang.

Notes to the Reader

Chinese and Japanese names have been presented in the traditional fashion, with the family name first, followed by the given name. Exceptions have been made for those who choose to adopt the Western system (e.g., Wen Fong).

In transliterating the Chinese language, the pinyin romanization system has been employed. Macrons have been used to indicate long vowels in Japanese names and words; commonly known city names (e.g., Tokyo, Osaka, Kyoto, and Kobe) have been given without macrons.

Throughout this publication, bibliographical sources have been cited in notes by author-date codes listed in the References accompanying each author's text. Entries in References have been ordered alphabetically by the author-date codes.

Rubbings of inscriptions are reproduced for all inscribed bronzes. Where two inscriptions are present, the top rubbing has been taken from the lid and the bottom from the body.

CHRONOLOGY

Xia dynasty	ca. 2100–ca.1600 BC
Shang dynasty	ca. 1600–ca.1050 BC
Zhou dynasty	ca. 1050–221 BC
Western Zhou dynasty	ca. 1050–ca.771 BC
Eastern Zhou dynasty	ca. 770–221 BC
Spring and Autumn period	ca. 770–ca.470 BC
Warring States period	ca. 470–221 BC
Qin dynasty	221–206 BC
Han dynasty	206 BC–220 AD
Western Han dynasty	206 BC–9 AD
Eastern Han dynasty	25–220
Three Kingdoms period	220–265
Six Dynasties period	220–589
Sui dynasty	581–618
Tang dynasty	618–907
Five dynasties	907–960
Liao dynasty	916–1125
Song dynasty	960–1279
Northern Song dynasty	960–1127
Southern Song dynasty	1127–1279
Jin dynasty	1115–1234
Yuan dynasty	1272–1368
Ming dynasty	1368–1644
Qing dynasty	1644–1911

For the purpose of describing the sequence of Bronze Age culture in China, the Shang dynasty is divided into three primary phases which take their names from archaeological sites:

Erlitou phase	early Shang	before 1600 BC
Erligang phase	middle Shang	ca. 1600–ca.1300 BC
Anyang phase	late Shang	ca. 1300–ca.1045 BC

Late Shang or the Anyang phase is further divided into four periods associated with the kings of Anyang and known by the ancient name of the Anyang area, which was called the "Wastes of Yin" or Yinxu.

	Rulers of Anyang
Yinxu period I	Pan Geng
	Yang Jia
Yinxu period II	Wu Ding
	Zu Jia
	Zu Geng
	Zu Ji
	Kang Ding
Yinxu period III	Lin Xin
	Wu Yi
	Wen (Wu) Ding
Yinxu period IV	Di Yi
	Di Xin

Chinese Ritual Bronzes:
Collections and Catalogues Outside China

Thomas Lawton

Western comprehension of ancient Chinese culture emerged slowly and grudgingly during the 19th and early 20th centuries. At the heart of the problem was the skepticism with which most Westerners regarded traditional Chinese scholarship, a dramatic example of which was the reluctance of Western specialists to accept an early date for Chinese ritual bronzes. James M. Menzies (1885–1957), who served in China as a missionary of the United Church of Canada from 1914 to 1932, described how, before the First World War, "no Western scholar would believe that there was bronze in the Shang dynasty.... Even...in 1930, when visiting the museums of North America and making rubbings of the inscriptions on Chinese bronzes I found no one willing to date any Chinese bronzes as old as Shang."[1] Determining the precise date of individual bronzes has remained a matter of prime concern for specialists up to the present.

In 1909, in a rare exception to the Western practice of assigning conservative dates (i.e., Zhou–Han dynasties) to virtually all Chinese bronzes, Otto Kümmel (1874–1952), the future Director of the Ostasiatische Kunstabteilung and Director General of the Berlin State Museum, described a vessel as dating from the Shang dynasty, without any qualification whatsoever.[2] Kümmel's willingness to be so decisive in an area of early Chinese art history that still remained uncertain is more easily understood in light of the 1903 catalogue for a Paris sale organized by Hayashi Tadamasa (1853–1906), where the same bronze is given a Shang date.[3] Kümmel later changed his mind about the attribution. In his *Chinesische Bronzen aus der Abteilung für Ostasiatische Kunst an den Staatlichen Museen Berlin* of 1928, he guardedly spoke of the bronze as "probably very early."[4]

Chinese specialists, by contrast, frequently assigned Shang dates to bronzes. For example, in an exhibition catalogue written by Chu Deyi (1871–1942) in 1924 for the international art dealer C.T. Loo (1880–1957), a number of pieces were identified as being of the Shang period.[5] Loo was, in every sense, a quiet, refined connoisseur-dealer and many of the finest Chinese bronzes now in Western collections passed through his hands.

The earliest non-Chinese illustrated bronze catalogue was published in Japan in 1848. That catalogue, *Shōsanrindō shoga bunbō zuroku*, illustrates the few Chinese bronzes then owned by the distinguished Japanese scholar and calligrapher Ichikawa Beian (1779–1858).[6] Ichikawa devoted his life to the study of Chinese culture and it is understandable that he acquired several ancient bronzes, as well as a number of Chinese jades, seals, and ceramics—probably from dealers in Nagasaki, the only Japanese port open to China and the West during the Tokugawa period (1615–1868).

One of the first Westerners to collect important early Chinese bronzes appears to have been Henri Cernuschi (1820–96). Although born in Italy, Cernuschi lived most of his life in France; his Chinese art collection is now housed in the Musée Cernuschi, Paris.[7]

In 1871 Henri Cernuschi took a trip around the world. While in the Far East during that ten-month trip he succumbed to the lure of bronze vessels, collecting approximately 1500 Chinese and Japanese examples. Most of those bronzes were Japanese and most of them date from later periods, but Cernuschi should be credited with having acquired several early Chinese bronzes that are extremely impressive.

Traveling together with Cernuschi was the French art critic Théodore Duret (1838–1927). In 1874, Duret published a small volume about the trip entitled *Voyages en*

Asie, which is important for what he says about art collecting at the time.[8] In Tokyo, where art treasures were flooding the market as a result of the social, political, and economic upheaval prompted by the Meiji Restoration of 1868, Duret described how Cernuschi acquired bronzes—some of them Chinese—by the hundreds and in lots. In Beijing, where the situation was very different, he reported that Cernuschi made purchases one by one and only after long bargaining and paying high prices. At first Cernuschi bought all the bronze vessels available in Beijing antique shops. Then, by increasing the amount he was willing to pay, he succeeded in acquiring rare examples from private Chinese collectors.[9]

Duret specifically cites basic reference works for the study of Chinese bronzes, noting as well that all of the Chinese dealers had copies of those texts which, he says, served as *vade mecum* (handbooks or guides). Duret mentions several of those so-called Chinese handbooks or guides and his comments are remarkable, especially when we remember that they were made in 1874 by someone who was, essentially, an informed tourist.

The first of the Chinese texts Duret mentions is the *Bogu tu*, which he describes as illustrating a large number of antiquities and having been published for the first time during the Song dynasty "around the year 1200 [actually the catalogue is believed to have been completed around 1123], and subsequently reprinted many times."[10] Duret also refers to the *Xiqing gujian*, describing that imposing compilation as a catalogue raisonné of the collection of ancient bronzes assembled by the Qianlong Emperor (1711–99; r.1736–96) and translating the title as "Record of Antiquities of Western Purity" (Mémoire des antiquités de la pureté occidentale).[11] Finally, Duret cites the *Jiguzhai zhongding yiqi kuanshi fatie* (published 1804), and refers to the author, Ruan Yuan (1764–1849), as most recently having served as a "viceroy of Canton."[12]

Several of the Chinese bronzes Cernuschi acquired in the Far East in 1871 date from the Shang and Zhou dynasties. Since Duret mentioned several important Chinese texts in *Voyages en Asie*—especially *Xiqing gujian*—it is surprising that he made no comment about two of the bronzes purchased by Cernuschi as having been illustrated in the Qing imperial catalogue.[13] Perhaps he and Cernuschi were unaware of that provenance, although it seems odd that the Chinese dealers did not point it out as a means of obtaining a higher price. Evidently the two bronzes were among the antiquities taken from the imperial collections during the looting of the Yuanmingyuan, or "Old Summer Palace," at the outskirts of Beijing by French and English troops in 1860.

Shortly after Cernuschi returned to France from his world tour, his entire collection of 1500 bronzes was shown in an international exposition at the Palais de l'Industrie, Paris from August 1873 to January 31, 1874, apparently the first time bronzes from the Qing imperial collection were publicly displayed outside China. Since Cernuschi does not seem to have been aware of their imperial provenance, the event went unnoticed by the press.

Two other bronzes from the Qing imperial collection now in the West left China under entirely different circumstances. In 1887, the Guangxu Emperor (1871–1908; r.1875–1908) presented a Shang dynasty bronze *gu* to Queen Victoria (1819–1901; r.1837–1901) to mark the fiftieth anniversary of her reign. Queen Victoria's anniversary bronze was displayed in the International Chinese Exhibition in London in 1935–36,[14] organized in celebration of the Silver Jubilee of Victoria's grandson, George V (1865–1936; r.1910–36), and again in London in 1951, in an exhibition of early Chinese bronzes organized by The Oriental Ceramic Society.[15]

In 1901, the Guangxu Emperor gave an Eastern Zhou period bronze ritual wine vessel, of the type *he*, to Kaiser Wilhelm II (1859–1941; r.1888–1918) of Germany.[16] The reason for the gift, from the Chinese point of view, was quite inauspicious. During the Boxer Rebellion of 1900, Baron Klemens von Ketteler, the German Minister to China, was murdered in Beijing by a government soldier. When foreign armies had put down the Boxers and signed a treaty with the Chinese government, the Chinese emperor sent his brother, Prince Chun, to Germany as his personal envoy to apologize for the murder and, as part of the royal visit, to present the bronze to the Kaiser.

During the 19th century, many Westerners were unable to differentiate between ancient Chinese bronzes and later copies—to say nothing of being able to determine whether bronzes had been made in China or Japan. A dramatic instance of how little expertise was available in the West in the 19th century is provided by the 150 bronzes acquired in 1876 by the Victoria and Albert Museum in London from the French art dealer Siegfried Bing (1838–1905).[17] Bing, who was instrumental in raising Western awareness of Far Eastern culture, identified all 150 bronzes as being Japanese, and for many years they were so labeled at the Victoria and Albert Museum.

As Western scholarship improved, however, some of the Victoria and Albert Museum pieces were identified as Chinese and, surprisingly, some have been recognized as archaic. For example, a square-sided ritual bronze *fangzun* is a genuine—if somewhat battered—Shang dynasty vessel.[18] Some of the vertical flanges that project beyond the rim of the vessel have been broken off and the animal heads that originally adorned the middle section of the vessel are missing. All that remains are the plugs to which those decorative heads were attached.

There are four square openings in the high flaring foot of the bronze *fangzun* in the Victoria and Albert Museum. Technological research during the past twenty years suggests those holes were caused by metal chaplets that held an interior clay core in place during the casting process. Chinese antique dealers passed on to unwary Western collectors a considerably more exotic explanation. According to their account, the holes were necessary so that if, during the ritual offering, the bronze should happen to be placed in such a position that the spirit to which the offering was made were imprisoned within the foot, the holes provided a means of escape; the spirit could still partake of the ritual foods and beverages.[19] While one has to admire the ingenuity of that explanation, even though it has no basis in fact, it evidently had great appeal to fledgling Western antique collectors. Looking at the Victoria and Albert Museum *fangzun* in the 1990s, it seems odd to think that anyone could have mistaken it for a Japanese piece. Yet we should remember that the attribution was made more than a century ago, at the very outset of serious Western interest in ancient Chinese culture.

The ancient Chinese bronzes in the Freer Gallery of Art are well known today, but the beginnings of that collection were remarkably tentative. Charles Lang Freer (1854–1919) purchased his first bronze, a *yu*, in 1894 from Rufus E. Moore (1840–1918), an art dealer with a shop at 33 Union Square in New York. Moore was one of the organizers of the American Art Galleries, located at Madison Square South in New York, where many important sales of Far Eastern art were held in the late 19th and 20th centuries. So little was known in 1894 that when Freer acquired the bronze he believed it to be a Japanese copy. Moreover, and this is even more astonishing, that attribution remained unchallenged until 1967, when the *yu* was correctly assigned to the Shang dynasty.[20] Almost a century after Freer acquired the *yu*—and enlightened by recent scholarship—we recognize the shape and decoration as being characteristic of pre-Conquest bronzes cast in the area of modern Shaanxi province in the 12th to 11th century BC.

Japanese art dealers played an important role in bringing many Chinese bronzes to Europe and the United States in the late 19th and early 20th century. While some of those bronzes had only recently become available to dealers because of events within the tottering Manchu regime, many of them which had been in Japanese collections for many years were believed to be of Japanese manufacture. It is possible the bronze *yu* purchased by Freer in 1894 had been in a Japanese collection before it came to the United States.

Among the most influential of Japanese art dealers, Yamanaka Sadajirō (1866–1936) deserves special praise.[21] On November 3, 1894, Yamanaka, then twenty-seven years old, realized his dream of traveling to the West when he boarded the steamship Empress of China bound for the United States. William Sturgis Bigelow (1850–1926), Edward Sylvester Morse (1838–1925), and Ernest F. Fenollosa (1853–1908), three of the most distinguished Americans to be seriously interested in Japanese culture, were instrumental in helping Yamanaka to open up his first shop at 4 West 27th Street in New York.

Those first years in New York were difficult, and as Yamanaka worked to establish a clientele, he experimented with all kinds of merchandise to attract customers. In later years he was fond of recalling the many Pekingese puppies and exotic goldfish that formed part of his initial inventory. The Pekingese sold well, the goldfish did not. By 1912 Yamanaka Sadajirō enjoyed a well-earned international reputation for honesty, service, and quality. He was also noted for his willingness to take risks. For instance, in 1912 Yamanaka purchased the bulk of the collection of the Manchu Prince Gong. When Yamanaka sold that collection at the American Art Galleries in New York in February 1913, the catalogue cover, decorated with an embossed gold design and red lettering, reflected the importance of the occasion.[22]

In 1914 Yamanaka organized another sale at the American Art Galleries. According to the title page of the catalogue, the objects comprised "the private collection of a Chinese nobleman of Tien-tsin and objects procured by the senior member of Messrs. Yamanaka & Co. and his staff during a recent visit to the ancient cities of China, some of which have heretofore been unexplored by foreigners."[23]

As we have seen, Japanese collectors and collections of Chinese art exerted great influence on Western connoisseurs like Charles Freer, and their initial reactions to ritual bronzes were tempered by Japanese connoisseurship. Several of the most outstanding Japanese connoisseurs included Kanō Jihei (1862–1951), a wealthy Osaka collector who assembled a small number of Chinese bronzes. In 1907 he published a two-volume catalogue entitled *Hakkakujō* that included a few of his ancient bronzes.[24] The initial two characters of the title, "*hakutsuru*" or "white crane," also appear in the name of the museum that now houses Kanō Jihei's collection, the Hakutsuru Bijutsukan in Kobe.

In 1911, Baron Sumitomo Kichizaemon (1864–1926), another wealthy Japanese connoisseur, published the first volume of his illustrated bronze collection, *Sen'oku seishō*; by 1916 all six volumes of that first version of Baron Sumitomo's Chinese bronze collection were available.[25] Subsequently expanded editions of the catalogue were published in 1919, 1921, and 1934 as the collection grew. In the introduction to the 1911 volume, Hamada Kōsaku (1881–1938) stated candidly that the study of Chinese ritual bronzes really had not yet begun in Japan. The first volumes of the Sumitomo catalogue appeared in a limited edition—meant for presentation—and were not offered for sale. They were difficult to obtain and those fortunate few, such as the English collector George Eumorfopoulos (1863–1939) to whom Baron Sumitomo presented copies, were understandably grateful. Japanese publications set high standards and the Sumitomo catalogue served as a model for Eumorfopoulos when he published his own collection in 1929.

In 1933 Umehara Sueji (1893–1983), at the behest of and with financial support from Yamanaka Sadajirō, completed his magisterial seven-volume work, *Ōbei shūcho Shina kodō seika*, recording the Chinese bronzes in European and American collections. In his preface, Umehara mentions that he examined Chinese bronzes in public and private collections during his three-year stay in Europe and the United States from early 1926 to the spring of 1929. Umehara also notes that Yamanaka gave him complete freedom in compiling data and writing the text, not interfering in any way, other than to ask that the volumes appear on schedule.

Following the positive reception of *Ōbei shūcho Shina kodō seika*, Yamanaka determined that there should be a companion publication for the ancient Chinese bronzes in Japanese collections, *Nihon shūcho Shina kodō seika*, and asked Umehara to take charge of the project. The six double-folio volumes, published in 1959–64 several decades after Yamanaka's death, were a fitting culmination of his lifelong support of scholarship.

Chinese scholars read the Japanese catalogues with avid interest and, in 1935, Rong Geng (1894–1983) published his study *Haiwai jijin tulu*, in which he discussed 158 Chinese bronze vessels in Japanese collections, many of them owned by Baron Sumitomo.[26] Rong Geng was prompted to compile his study because many ancient bronzes of exceptional quality had left China, making it difficult to see them and, equally important, the volumes in which those ritual vessels were illustrated were expensive and not easily obtainable.

The Sumitomo collection of Chinese bronzes was and remains the most important in Japan. Baron Sumitomo wrote about his collecting in *Sumitomo Shunsui*;[27] according to that account he purchased his first Chinese bronze, a vase of the type favored in Japan for use in the tea ceremony, in 1896 or 1897. By the turn of the century, however, Baron Sumitomo had expanded his holdings to include older and more important pieces.

To the best of my knowledge, the first public exhibition of ancient Chinese ritual bronzes in the United States took place in 1916 at the Metropolitan Museum of Art in New York City.[28] Ironically, the theme of that exhibition was Chinese pottery and sculpture—the bronzes were included at the last moment, almost as an afterthought. In retrospect, the decision to display those few bronzes, each one of outstanding quality, marked a milestone in the history of American understanding and appreciation of early Chinese culture.

All of the bronzes shown at the Metropolitan in 1916 are now in the Freer Gallery of Art. They were acquired in China by Marcel Bing (1875–1921), a young French art dealer.[29] It was Marcel Bing's father, Siegfried Bing, who sold the 150 so-called "Japanese" bronzes to the Victoria and Albert Museum in 1876. Marcel had been trained as a jeweler and, while he pursued that career in the family business, his first love was Chinese art. Unlike his father, Marcel Bing appears to have had an intuitive response to quality in his selection of Chinese antiquities.

During the First World War, Marcel Bing fought with the Allies against the German forces of Kaiser Wilhelm II. As a soldier he was unable to manage the family business and soon was in dire need of money. In 1915 Marcel Bing offered his private collection of Chinese bronzes to Charles Freer for $100,000. Freer discussed the collection with Eugene Meyer, Jr. (1875–1957) and his wife, Agnes Ernst Meyer (1887–1970). On many occasions Freer visited art dealers together with the Meyers, sharing opinions and, from time to time, making purchases jointly. Whenever they made joint purchases, Mr. and Mrs. Meyer insisted that Freer should have first choice, since his collection was intended for a national museum in Washington, D.C. Further, the Meyers told Freer that they intended to bequeath their Asian art collection to his museum.[30]

In 1915, on Freer's recommendation, Mr. and Mrs. Meyer agreed to join him in purchasing Marcel Bing's Chinese collection. In one fell swoop they acquired eleven superb Chinese bronzes. It was one of the best purchases they ever made and Freer had come a long way from his 1894 purchase of what he believed to be a Japanese bronze *yu*. Freer and the Meyers were so pleased with their new acquisitions that they offered to show them at the Metropolitan Museum, together with some of their Chinese pottery and sculpture, in 1916.

The English collector George Eumorfopoulos was one of the Western pioneers in supporting the study of Chinese ritual bronze vessels. In 1929, ten years before his death, Eumorfopoulos published a catalogue of his bronze holdings in two editions: one limited to 560 numbered copies, the other a deluxe edition of twenty-five copies.[31] The volume was one of eleven—each covering a wide range of Chinese antiquities—that set new standards for Western scholarship on Chinese art. Aside from the high quality of the bronzes acquired by Eumorfopoulos, the scholarly text for the catalogue prepared by Sir Perceval Yetts (1878–1957), a medical doctor who went to China for the first time in 1908, reflects a new mastery of Chinese and Japanese sources. Eumorfopoulos also wrote an introduction that is particularly informative. He identified a bronze *you* in his collection as being indicative of a change in Western appreciation of ancient Chinese ritual vessels. According to Eumorfopoulos, when the *you* arrived in London in 1910, it came as a revelation to British collectors, who were unaccustomed to the standards it set in craftsmanship.[32] The sale of the bronze for one thousand pounds—a considerable sum at the time—also set a record for the price paid for a Chinese bronze.

In the early decades of the 20th century, when interest in all aspects of Japanese art was predominant, a series of exhibitions signaled the gradual emergence of Chinese art as a subject of international significance. In September and October 1925, the Stedelijk Museum in Amsterdam held an Exhibition of Chinese Art that included several important Chinese bronzes.[33] In the exhibition catalogue the bronzes were grouped under the broad category, "Plastic Art (Figures and Ornamental Works)." An introductory essay by T.B. Roorda, President of the Society of Friends of Asiatic Art, provides general information, together with a brief discussion of some specific terms, such as *leiwen* and *taotie*, while the individual bronzes were dated "Zhou" and "Pre-Han."

The *Chinesische Kunst* exhibition, which opened in the Preussischen Akademie der Künste in Berlin on January 12, 1929, was envisioned as providing Europeans with their first opportunity to appreciate the full range and development of China's artistic traditions.[34] By any standards, the Berlin exhibition was organized on a large scale; the catalogue contained 1272 entries. All of the ritual bronzes included in that display were dated to the Zhou or Han periods.[35]

At the suggestion of the Swedish Crown Prince, Gustav Adolf (1882–1973), in September 1933 the Museum of Far Eastern Antiquities in Stockholm organized The Exhibition of Early Chinese Bronzes on the occasion of the thirteenth International Congress in the History of Art.[36] Scholars, collectors, and connoisseurs who attended the congress discussed a number of questions relating to Chinese bronzes that still remain vital.[37] Again, at the behest of the Swedish Crown Prince, officials charged with installing the bronzes devised a "novel arrangement in an effort to establish a chronology of Chinese bronze art in greater detail than had hitherto been attempted."[38] While the opinions offered by international scholars in 1933 were extremely tentative, they helped to lay the groundwork for research that still occupies bronze specialists in China and the West.

The following year Georges Salles (1889–1966), conservateur of the Département des Arts Asiatiques in the Louvre, organized an exhibition entitled *Bronzes chinois, des dynasties Tcheou, Ts'in, et Han* in the Musée de l'Orangerie in Paris.[39] The first fifteen bronzes in the catalogue were the so-called "Liyu" bronzes acquired shortly after they were unearthed in Shanxi province by Leon Wannieck, a French businessman.[40] The remaining pieces were drawn from French, Swedish, and English collections.

All of the earlier European exhibitions culminated in the International Exhibition held in London in 1935–36. With remarkable energy, the organizers of the London exhibition assembled more than 3000 objects for display in the Royal Academy of Arts. Building on the advances achieved in earlier presentations, the London exhibition, catalogue, lectures, and associated publications marked a turning point in Western understanding of Chinese art, including ancient ritual bronzes. The group of Chinese bronzes on display, 108 of which were loaned by the Chinese government, with equally generous loans by collectors in Japan, Europe, and the United States, included some of the most impressive examples then known and indicated the enormous variety of shapes, motifs, and styles. According to one person who accompanied the objects loaned by the Chinese government, the two bronzes that attracted the most attention during the exhibition were the owl *zun* and the double ram *zun*, both then owned by George Eumorfopoulos and now in the Victoria and Albert Museum and the British Museum respectively.[41] Dating of the bronzes in the exhibition catalogue essentially followed the six-period scheme established by the Chinese Organizing Committee.[42]

In Paris, René Grousset (1885–1952), director of the Musée Cernuschi, organized a bronze exhibition drawn from French and Swedish collections, which opened in May 1937.[43] The purpose of the exhibition was to illustrate the development of ancient Chinese bronzes according to the principles set forth by the Swedish scholars Johan Gunnar Andersson (1874–1960), Bernhard Karlgren (1889–1978), and Orvar Karlbeck (1879–1967); it also included references to the research on the relationship between neolithic pottery decoration and that on bronzes, as proposed by Max Loehr (1903–88).[44]

In the United States, one of the immediate results of the London exhibition was a show of Chinese bronzes from the Shang through the Tang dynasty held at the Metropolitan Museum in New York from October through November of 1938.[45] Alan Priest (1898–1969), the charming, completely unpredictable curator of Far Eastern Art at the Metropolitan, organized that exhibition. In his introduction to the catalogue, Priest wrote, "The present exhibition...is drawn from but one source—American collections—but those who saw the great exhibition of Chinese art in London in 1935 will recognize that there are in this country collections comparable to those drawn upon on that occasion from the whole world." In his article for the Museum *Bulletin*, Priest waxed even more ecstatic, proclaiming, "It is such a galaxy of bronzes as has never been seen together."[46]

Priest had good reason to sound self-satisfied. He had succeeded beyond his wildest dreams in assembling a truly impressive group of ancient bronze ritual vessels. Twenty-three of those pieces were from the Pillsbury collection in Minneapolis; ten were loaned by the Nelson-Atkins Museum in Kansas City; while others came from the Cleveland Museum of Art, and the Field Museum in Chicago.

From the Metropolitan Museum's own holdings, the centerpiece, of course, was the famous "Duanfang altar," which the museum acquired from the widow of Duanfang (1861–1911) in 1924.[47] According to one Chinese source, the sale price was "more than 200,000" silver dollars.[48] Regardless of the cost, the so-called "Duanfang altar" remains one of the most important groups of early Western Zhou period Chinese ritual bronzes in the world.

Some of the other Chinese bronzes from the Metropolitan's own holdings were part of the bequest of Mrs. Otto Kahn (1876–1949). Mrs. Kahn and her husband were generous supporters of the New York cultural scene, and when she died in 1949, Mrs. Kahn left fifteen Chinese ritual bronzes—then valued at $55,100—to the museum.[49] One of Mrs. Kahn's bronzes, an early Western Zhou period ritual *you*, was among the pieces found in China by Dr. Jörg Trübner (1901–30).[50] It is difficult to speak of Trübner, even briefly, without using superlatives. He had remarkable taste and, even though he died tragically young while in China, the objects Trübner acquired and made available to private and public collections speak eloquently of his enviable sense of connoisseurship. Even the Chinese antiquarians who met him remarked on Trübner's extraordinary understanding of Chinese antiquities.[51]

The 1938 exhibition at the Metropolitan Museum was important because it acknowledged the emphasis given to ancient Chinese bronzes in the United States. Scholarship, as manifested in the catalogue, still lagged behind comparable investigations in China and Japan, but it was clear that at least some Americans were beginning to pay serious attention to the subject.

The following year W. Perceval Yetts published his catalogue of the Cull Chinese bronzes in a limited edition of 350 copies.[52] The bronzes, assembled by Anders Eric Knös Cull (1878–1968) and James K. Cull, English-born bankers of Swedish ancestry, were presented with admirable detail by Yetts. It is noteworthy for an understanding of changes in ownership that, at the time of the 1935–36 London exhibition, the famous Zhao Meng *hu*, the so-called "Cull *hu*" now in the British Museum, were owned by Edgar Worch (1880–1972).

The Cull catalogue is distinguished by its handsome design and admirable scholarship. As one scholar noted, Yetts was "a true archaeologist and not a mere historian of Chinese art, [who] has fully realized the fundamental importance of this branch of sinology, which is often sadly underrated and neglected by European scholars."[53] In spite of Yetts's care in examining matters of provenance and inscriptions, he conservatively assigned the individual bronzes to "phases," a chronological and stylistic sequence he had introduced in the Eumorfopoulos catalogue and which was gradually refined by later scholars during succeeding decades.

The year 1939 also saw the publication of the catalogue of twenty bronzes acquired by Dr. Oskar Paul Trautmann (1877–1950) when he served as German ambassador to China from 1935 to 1938. Those were critical years for diplomatic relationships between China and Germany, culminating with the fall of the capital of Nanjing in 1937. Trautmann pursued intense negotiations with Chiang Kai-shek and the Nationalist Government before leaving China in June 1938.

The Trautmann catalogue, *Frühe chinesische Bronzen aus der Sammlung Oskar Trautmann*, prepared by Gustav Ecke (1896–1971), presents each of the pieces in excellent black-and-white collotype illustrations, supplemented by rubbings of decoration and inscriptions. The slender volume has a traditional Chinese blue-cloth case.[54] Some of the Trautmann bronzes had passed through the hands of the well-known Beijing art dealer Huang Jun, while the Chinese title for the catalogue was selected by Rong Geng.

Sammlung Lochow: Chinesische Bronzen, Ecke's research on a portion of the bronzes assembled by Hans Jürgen von Lochow (1902–89) while in China from 1934 to 1939 as an advisor to the Chinese government on railroad construction, was published in 1943. The second volume, with a text written by the collector, was published the following year.[55] When von Lochow, a German national who had resided in China for many years, returned to Germany in 1950, Chinese authorities permitted him to take his Chinese bronzes, with one exception. The exception was a large bronze *nao* bell that provided the Chinese name for von Lochow's studio, Naozhai, and was given pride of place as the first and most detailed entry in the initial volume of the von Lochow catalogue. Determining the bell to be a National Treasure, Chinese authorities insisted it remain in China.[56]

These two German-language catalogues are noteworthy because they were printed in China, presented collections that were formed while their owners resided in China, and were prepared by Gustav Ecke, a fellow German national who was a longtime resident in China. Werner Jannings (d. ca.1959), yet another German businessman who lived in China for many years, also assembled a collection of some 240 bronze antiquities that was acquired by the Chinese government on January 22, 1946, and displayed in the Zhongcuigong within the Palace Museum.[57]

If 1935–36 was a turning point in Western studies of Chinese bronzes, and 1938 provided recognition of the riches of American holdings of ancient bronzes, 1946 was a banner year for scholarly achievements in the United States. In that year catalogues of both the Buckingham[58] and Freer[59] collections were published: the first scholarly catalogues of specific American collections of Chinese bronzes to appear in print. They were and still remain impressive volumes. Some idea of how the prices of museum catalogues have changed over the years can be gained from the fact that in 1946 the Buckingham catalogue cost $7.50; the Freer volume sold for an even more modest price of $5.00.

The Buckingham collection of Chinese bronzes was acquired and presented to the Art Institute of Chicago by Kate Sturges Buckingham (1858–1939) in memory of her sister, Lucy Maud Buckingham (d.1920). As officials at the Art Institute planned the catalogue, they turned to Charles Fabens Kelley (1885–1960) and Chen Mengjia (1911–66), two of the most esteemed of contemporary scholars. Charles Fabens Kelley was a member of the staff of the Art Institute for thirty-two years. When Kelley retired he was Assistant Director and Curator of Oriental Art. Chen Mengjia was a leading Chinese antiquarian who was in the United States from 1944 until 1947, teaching Chinese paleography at the University of Chicago. The Art Institute took advantage of Chen's being in America to have him collaborate with Kelley on the Buckingham collection.

The first catalogue of the Freer Chinese bronzes, published in 1946, was dedicated to John Ellerton Lodge (1876–1942). The fifty-six bronzes in the catalogue were selected from among those acquired during Lodge's directorship; they remain the most important group of bronzes in the Freer Gallery.[60] Even in such a scholarly text there are glimpses of Lodge's quiet humor. Of one early Western Zhou bronze *gui* Lodge wrote, "When the dealer brought this vessel to the Gallery in the Autumn of 1937, he did not know (or would not tell) where and when it had been excavated; but he said that he had got it in China, and this *could* be true."[61]

Other catalogues of Chinese bronzes have been published through the years. The catalogue of the Alfred F. Pillsbury (1869–1950) holdings, written by Bernhard Karlgren, presents a collection that the owner bequeathed to the Minneapolis Institute of Arts, together with his other holdings, in 1950.[62] Pillsbury purchased almost all of his bronzes from C.T. Loo and the high quality of the pieces reflects Loo's emphasis on offering the finest pieces available to his customers.

The Willem van Heusden collection, since dispersed, was compiled by the owner for *Ancient Chinese Bronzes of the Shang and Chou Dynasties*, published privately in an edition of one thousand.[63] Van Heusden introduced his collection with a general summary of early Chinese history, and a discussion of ritual functions and casting techniques. He analyzed the bronze inscriptions and acknowledged that several pieces were formerly owned by Liang Shangchun.[64]

Early Chinese Bronzes, the catalogue published in 1956 by the City Art Museum of Saint Louis, has a text by J. Edward Kidder, Jr.[65] Most of the bronzes included in that publication were gifts from J. Lionberger Davis (1878–1973), a distinguished banker and civic leader. Other pieces were given in memory of Dr. J. William Beckmann (d.1954), and there were several museum accessions as well. Considerably earlier than those generous donations, beginning in 1916, the museum purchased a few impressive vessels from Rufus E. Moore, the same New York art dealer who sold Charles Freer his first archaic Chinese bronze in 1894. An early Western Zhou *gui* (No. 26; 203:1916), one of the bronzes acquired by the museum in 1916, is indicative of the types of Chinese antiquities that were available on the international market at the beginning of the century. The distinctive coiled-dragon decoration on the *gui* appears on bronzes made at the end of the Shang and beginning of Western Zhou.[66] A description of the *gui*, written soon after it was acquired by the museum, captures the prevailing Western attitude toward Chinese antiquities:

What gives these bronzes their greatest attraction, aside from the archaeological point of view, is the remarkable color the metal assumes; the reds, grays, and yellows, the malachite green incrustations, a blue approaching that of ultramarine, where a lapse in the polished surface permits the erosion of the copper. The thought of the designer and the labor that produced the elaborately chiselled surface furnish a fitting foundation for the crowning handiwork of Nature.[67]

Among the many impressive bronzes donated by J. Lionberger Davis, the Xiaochen Yi *jia* remains one of the most important (No. 20; 221:1950). The earliest reference to the bronze and its twenty-six character inscription, is in *Taozhai jijin lu*, the catalogue of the extensive holdings of the collector Duanfang.[68] The inscription, one of the few Shang dynasty examples of such length, reads: "On the day *gui si* the king awarded the Xiaochen Yi ten strings of cowries, which he used to make this sacred vessel for Mu Gui. It was in the king's sixth *si*, during the *yong* cycle, in the fourth month. *Ya Yi*."[69] Following Duanfang's assassination, the Xiaochen Yi *jia* was obtained by Rong Hou (1875–?), and was included in his catalogue *Kankarō kikkinzu*.[70] By 1935 the bronze was in the collection of Wang Chen (1909–36) and is published in *Shierjia jijin lu*.[71]

A small bronze *zhi*, or beaker, in The Saint Louis Art Museum—another gift from J. Lionberger Davis—is remarkable for its unusual decoration (No. 23; 215:1950). The peacock-like plumage of the birds combined with *taotie* faces, and the spined vertical flanges articulated by a chevron pattern, are characteristic features of bronzes cast during the latter years of the Shang dynasty.[72] Kidder described this *zhi* as "the most extraordinary bronze in the collection; its decoration and the interpretation of it are quite unique." [73] When he commented on the Saint Louis *zhi* in 1968, Max Loehr referred to a similar example that was available on the Beijing antique market prior to 1945.[74] A spurious bronze *zhi*, now in the Palace Museum, Beijing, might be the one mentioned.[75]

Chen Mengjia, who made a comprehensive survey of the Chinese ritual bronzes in American public and private collections when he was in the United States from 1944 to 1947, also provides a subjective analysis of developments during the first half of the 20th century.[76] Even though some of Chen's statements are inaccurate, they still remain important since they reflect a prevailing Chinese point of view. For example, Chen neglected the early acquisitions made during the late years of the 19th century by people like Charles Freer, and dates the first phase of American interest in Chinese bronzes to 1910–27, when businessmen or missionaries who had lived in China acquired examples and brought them back to the United States. More informative is his statement that many of those vessels were forgeries, with most of the others being ordinary *shoukeng*, or bronzes that had been unearthed many years earlier and treasured by Chinese connoisseurs who added layers of wax, varnish, or pigments to protect the metal surfaces, and, in so doing, darkened and polished the original appearance of the bronzes.

According to Chen Mengjia, it was during his first phase that Japanese dealers, such as Yamanaka Sadajirō, opened antique shops in New York and other American cities, with their agents purchasing bronzes from all parts of China. At that time, according to Chen, antique dealers from southern China also sent representatives to the United States to offer good, bad, and spurious bronzes to American customers. He refers specifically to a group of gilt bronze vessels that had been fabricated in Suzhou and sold in the United States.[77]

Chen Mengjia also refers to the so-called Duanfang Baoji set of bronzes and to the four bronzes of the Marquis of Qi in the Metropolitan Museum as being the most important pieces to reach the United States during those years. John C. Ferguson (1866–1945) played a key role in the negotiations for them. The Marquis of Qi bronzes from the collection of Shengyu (1850–1900) were secured by Ferguson for the Metropolitan Museum in 1912.[78] The set was said to have been discovered near Yi Xian, Hebei province in 1893.

In the decade from 1927 to 1937, which Chen Mengjia designates as the second phase of his scenario, the center of the international antiquities market moved from Europe to the United States, a transition that had been underway since the First World War, when Paris gradually relinquished its position as the major city for the acquisition of Asian art. Chen's third phase extended from 1937 to 1946, ending with the Second World War. During those years the finest Chinese bronzes came to the United States as Americans were able and willing to acquire the most expensive pieces.

According to Chen, Chinese art dealers were quick to note the shift in the center of the Chinese antiquities market. They developed their own jargon when designating specific bronzes for potential markets at home and abroad. For example, *yangzhuang* (foreign business) was the term they used for those bronzes that would appeal to foreign buyers. They noted, as well, that foreign collectors preferred *shengkeng* pieces, or those bronzes that had been excavated recently and still retained the patina formed during centuries of burial, as opposed to the dark shiny *shoukeng* surfaces that were a standard feature of traditional Chinese connoisseurship of Chinese ritual bronzes. In their efforts to provide bronzes for the international market, Chinese dealers became adept at removing the wax, paint, etc. from *shoukeng* vessels so that they would resemble the more desirable *shengkeng* bronzes. Chinese art dealers occasionally added pigments or experimented with acid to enhance the color of some bronzes in their attempt to achieve the attractive patinas admired by their foreign customers.

Since foreigners demanded bronzes that were completely intact, even though many bronzes were damaged during excavation, Chinese dealers skillfully repaired vessels and, at times, those repairs involved extensive reconstruction.[79] Western collectors also insisted on bronzes with fine decoration, as well as animal-shaped vessels, which prompted dealers to cater to their preference with bronzes unearthed at Anyang and from Warring States period tombs.

Given the categories identified by Chinese dealers, it is understandable that so many of the bronzes in the United States date from the Shang and Warring States periods, that many of those bronzes had gold and silver inlay decoration, and that they included three-dimensional animal and figural pieces. Correspondingly, Chinese dealers reserved different categories of bronzes for the local market: those bronzes with ordinary decoration but with important inscriptions—always a priority for Chinese connoisseurs; those pieces with good shapes, albeit not well cast; and the more usual Western Zhou and Spring and Autumn period pieces.

A giant step in the definition of the style of early bronzes came in 1953, when Max Loehr published his article on the bronze styles of the Anyang period.[80] In one brilliant stroke, basing his arguments on theoretical analysis, Loehr brought order to a stylistic sequence that previously had seemed cluttered and confused. It took some daring for Loehr to challenge the major specialists of the time, some of whom refused to acknowledge his argument's validity long after it had been accepted by virtually everyone else.

Archaeological excavations played a crucial role in demonstrating the accuracy of Loehr's theory. In 1968, he was able to amplify his views regarding style and chronology—extending through Shang and Zhou—in the exhibition *Ritual Vessels of Bronze Age China*, held at the Asia House Gallery.[81] The value of Loehr's article and his catalogue, as well as their weakness, lies in their presentation of neat and tidy stylistic sequences. Viewed in light of what is known almost a quarter of a century later, we must add that the casting of ritual bronzes in ancient China, their stylistic relationships, and their geographical distribution were far more complex than Loehr anticipated.

Research on ancient Chinese bronzes during the past few decades, as in the past, has depended heavily upon archaeological finds in the People's Republic of China. Initially visitors brought back reports of new finds and photographs—some of them frustratingly blurred—gleaned from Chinese periodicals that helped to fill some of the gaps in our knowledge. Then, in 1980, the Metropolitan Museum organized *The Great Bronze Age of China* exhibition. It was a spectacular presentation, supplemented by an outstanding catalogue written by distinguished scholars.[82] Once again we were indebted to the Metropolitan Museum for establishing new guidelines; the museum should be given credit for its many contributions, beginning as early as 1916, to the study of ancient Chinese bronzes.[83] Anyone who saw the 1980 exhibition or who has read the catalogue, which remains a basic reference work, has to be aware of the shift in the balance of scholarly contributions. The comprehensive analyses of style, of chronological sequence, and of bronze inscriptions all were eloquent reminders of how much Western scholars could contribute. Some of the Western concepts were so troubling to Chinese colleagues that they felt compelled to include their own views in a separate summary added at the end of the catalogue. Western scholarship had come a very long way from the 19th century, when even the distinction between Chinese and Japanese bronzes was an insuperable problem.

With the publication of *The Great Bronze Age of China* it was clear that Chinese and Western scholars had attained mutual status, that they could be described as colleagues of equal standing. The magisterial catalogues supported by Dr. Arthur M. Sackler (1913–87) established even higher scholarly standards. In 1983 Dr. Sackler published a preliminary study of a selection of bronzes from his collection in a catalogue entitled *Art from Ritual: Ancient Chinese Bronze Vessels from the Arthur M. Sackler Collections.*[84] That catalogue, written by Dawn Ho Delbanco, was only a prelude, however, to a much more ambitious scholarly project funded by Dr. Sackler that resulted in the publication of Robert W. Bagley's study of Shang dynasty bronzes in 1987,[85] and Jessica Rawson's study of Western Zhou bronzes in 1990.[86] The third and final installment of this project—a comprehensive examination of Eastern Zhou bronzes by Jenny F. So—appeared in 1995.[87]

The Sackler bronze catalogues should be regarded as milestones in the definition of our present knowledge about Chinese ritual bronzes in the West. To appreciate the magnitude of that achievement, one need only look back at those scholars and connoisseurs who first grappled with the basic questions relating to the shapes, motifs, and inscriptions that are so critical to a proper perception of these Chinese cultural monuments.

Our understanding of Chinese bronzes continues to develop, and there is every reason to believe that the high level of research that resulted in innovative publications such as *The Great Bronze Age of China* and the Sackler volumes will continue into the 21st century. Scholars in China and in the West are fully aware that many significant problems remain to be resolved during the next century, including such vexing questions as possible relationships between the imagery found on the lithic ritual objects of China's various Neolithic cultures and the motifs on bronzes dating from the early Shang dynasty. Equally compelling, not to say complex, is the question of how widespread was China's bronze industry during the Shang dynasty and how influential were the foundries at Anyang.

The variety, sophistication, and geographical breadth of Shang cultures, facets of ancient China's bronze culture that have been highlighted dramatically by recent finds at Sanxingdui, Guanghan, in Sichuan province,[88] and at Dayangzhou, Xin'gan, Jiangxi province,[89] have alerted scholars to hitherto unexplored, unanticipated possibilities, suggesting among other things that Anyang may have been dependent upon decorative motifs and technological innovations from the south. The importance of southern Chinese cultural achievements has the potential to modify our vision of ancient China's bronze industry and to lead scholars into totally unanticipated directions. The possibilities for new discoveries during the 21st century are exciting to contemplate.

NOTES

Some of the ideas developed in this chapter were initially explored in a paper, "Western Catalogues of Chinese Ritual Bronzes," presented at a conference celebrating the opening of the Arthur M. Sackler Museum of Art and Archaeology, Beijing, in May 1993.

1. Menzies 1940.

2. Lehnert 1909, 2: color pl. facing 728.

3. *Collection Hayashi* 1903, entry 998; sale held 16–21 February. A footnote cautions potential buyers that the Shang date is traditional and refers to a time that is "somewhat prehistoric."

4. See Kümmel 1928, 8, pl. 3, where the bronze is illustrated together with an identical piece in the Berlin collection, both of which Kümmel says formerly were owned by the Japanese connoisseur Machida Hisanari (1838–97).

5. Tch'ou 1924.

6. Ichikawa 1848, vol. 8. Complete in ten volumes, this work offers illustrations of the Zhou and Han dynasty bronzes in volume 8. There are only a few bronzes and some of those appear to be forgeries. For a discussion of early Japanese collectors of Chinese ancient bronzes, see Mizuno 1968, 21–23; and Hayashi 1984, 1:17–21.

7. The Italian-born Cernuschi was an ardent libertarian; following the Lombard revolution of 1848 he fled to France where he became a prominent politician, economist, and banker. See Elisseeff 1977; and Bobot 1983. Also see *Orientations* 1992:23(8) for a series of articles about the Musée Cernuschi.

8. Duret 1874.

9. Ibid., 20–21; Elisseeff 1977, preface.

10. Since there is no preface to *Bogu tu*, the author(s), circumstances, and precise date of the compilation of the Song imperial catalogue remain uncertain. Traditionally the task of compiling the thirty-*juan* catalogue is said to have begun in 1107 and to have been completed at some time after 1123. The number of ancient bronzes assembled by Emperor Huizong (r.1100–26) also is uncertain; while only 839 pieces are recorded in *Bogu tu*, comments by contemporaries suggest that the Song emperor eventually owned many more than that. In 1851, almost twenty years earlier than Duret's comments, Peter Perring Thoms published a translation of a portion of *Bogu tu* entitled *A Dissertation on the Ancient Chinese Vases of the Shang Dynasty, from 1743 to 1496 B.C.* In contrast to the comments by Duret, Thoms prepared his translation as an intellectual exercise. Thoms also made the mistake of having the Song illustrations redrawn by a contemporary Chinese artist. There is no indication that Thoms had ever seen an archaic Chinese ritual bronze and a careful reading of his translation reveals that Thoms believed the vessels to be made of gold.

11. For a comprehensive discussion of the titles of Qianlong's bronze catalogues, see Liu 1989, 1–2.

12. Actually, Ruan Yuan was Governor-general of Guangdong and Guangxi from 1817 to 1826, with his official residence in Canton. In 1797 Ruan Yuan reprinted the Song-dynasty compendium *Lidai zhongding yiqi kuanshi* compiled by Xue Shanggong. It is clear from his preface to *Jiguzhai zhongding yiqi kuanshi* that Ruan Yuan intended his compilation to be a continuation of Xue Shanggong's magisterial publication. There is, nonetheless, an obvious sense of satisfaction in his statement that while Xue Shanggong's catalogue contained 493 inscriptions, *Jiguzhai zhongding yiqi kuanshi* presents 551.

13. Elisseeff 1977, entries 29, 52.

14. *International Exhibition* 1935–36, 200, entry 2342; 215, fig. 2342.

15. "Catalogue of the Exhibition" 1953, 79, entry 1. The date of this exhibition coincided with the centenary anniversary of the publication of Thoms's *Dissertation on the Ancient Chinese Vases*.

16. Münsterberg 1924, 2:141, fig. 229.

17. A detailed study of Siegfried Bing and his career is provided in Weisberg 1986.

18. Koop 1924, pl. 6. For a review of Koop's book see Hentze 1927.

19. Noted in Koop 1924, 41.

20. Pope, et al. 1967, 346–49, pl. 62.

21. For biographical information about Yamanaka Sadajirō see *Yamanaka* 1939; Kuwabara 1967; Segi 1991; Lawton 1995.

22. *Illustrated Catalogue* 1913. The sale was held on February 28.

23. *Illustrated Catalogue* 1914. The sale was held on January 29–30.

24. *Hakkakujō* 1907.

25. *Sen'oku seishō* 1911–16. Six large volumes of plates, many in color, with descriptions in Japanese and English by Professors Hamada Kōsaku (1881–1938) and Harada Yoshito (b.1885).

26. Rong 1935.

27. Sumitomo 1955.

28. *Catalogue of an Exhibition* 1916.

29. Details of Marcel Bing's career and the negotiations for his collection of Chinese art appear in Luo Tan (Lawton) 1992.

30. *Eugene and Agnes E. Meyer* 1971.

31. Yetts 1929. For reviews, see Fischer 1935; Morgan 1929; Pelliot 1930; Sirén 1933.

32. Loo 1940 states that the bronze *you* arrived in England in 1910. According to Eumorfopoulos (p. v), the *you* was purchased by William Cleverly Alexander as a result "of a 'deal' in which a famille-noire vase he had been holding for some time played an important part. It speaks volumes for Mr. Alexander's courage and flair that he decided to part with what had always been regarded as 'gilt-edged security' in the collecting world for something that the true value of which had not been properly appraised." Two or three years after Alexander's death, Eumorfopoulos was able to acquire the *you* for his own collection, "at a considerably enhanced price" (p. vi).

33. *Exhibition of Chinese Art* 1925. Also see Visser 1926.

34. *Ausstellung chinesischer Kunst* 1929; Kümmel 1930a.

35. For a poetic reaction to the ritual bronzes in the 1929 Berlin exhibition, see Ayscough 1929.

36. "Exhibition of Early Chinese Bronzes" 1934.

37. Ibid., 84–85.

38. Ibid., 81. Orvar Karlbeck and Nils Palmgren developed five stylistic divisions as guides in dating ancient bronzes: (1) Yin, 11th century BC and earlier; (2) Yin-Zhou, 11th and 10th centuries BC; (3) Middle Zhou, 9th–7th century BC; (4) Huai, 7th or 6th century–3rd century BC; (5) Han, BC 206–220 AD and somewhat later. The term "Huai," suggested by René Grousset (ibid., 86), defined a bronze style that flourished in the Huai River valley during the later half of the Zhou and during the Qin dynasty. The much-criticized term finally fell into disuse. Bronzes of the first four periods are elaborated upon in Karlgren 1936.

39. Salles 1934a. A special issue of *Revue des arts asiatiques* 1934:8(3), was devoted to articles about the exhibition.

40. Salles 1934b.

41. Niu and Wang 1989.

42. (1) Shang, prior to 1070 BC; (2) Western Zhou, ca. 1070–722 BC; (3) Spring and Autumn Annals, ca. 722–481 BC; (4) Warring States, ca. 481–205 BC; (5) Han, ca. BC 206–220 AD; and (6) Post Han. In lectures presented during the London Chinese exhibition, Sir Perceval Yetts suggested a more general

chronological scheme: (1) First Phase (Shang-Yin and Early Zhou), 13th–11th century BC; (2) Second Phase (Zhou), 10th–6th century BC; (3) Third Phase (Zhou), 6th century BC to end of Zhou.

43. Grousset 1937.

44. Loehr 1936.

45. *Chinese Bronzes* 1938.

46. Priest 1938.

47. For a study of Duanfang, his art collections, and his catalogues, see Lawton 1997.

48. See Reitz 1924, where, regarding the date of the bronzes, the author states, "they date from the Chou period (1122–256 BC), and probably from the later part"; and Rong 1941, 168. For detailed discussions of the circumstances relating to the so-called "Duanfang altar," see Okada 1953; Umehara 1933; Umehara 1959.

49. Lippe 1950. For the bronzes' valuation, see *The New York Times*, 16 September 1953, sec. L, 68.

50. Kümmel 1930b, pls. 14–15; Chase 1991, 53, color pl. 3.

51. For a brief evaluation of Jörg Trübner's career, see Kümmel 1930b, 5–9. Trübner's contribution to the study of Chinese bronzes includes his book *Yu und Kuang: Zum Typologie der chinesischen Bronzen* (Trübner 1929).

52. Yetts 1939. For detailed reviews of the catalogue, see Erkes 1940; Kümmel 1941; Loehr 1948.

53. Erkes 1940.

54. Ecke 1939.

55. *Sammlung Lochow* 1943 and 1944. The major portion of the von Lochow bronzes is now housed in the Museum für Ostasiatische Kunst, Cologne.

56. Now in the Palace Museum, Beijing, the *nao* bell is reproduced in Weng and Yang 1982.

57. See Loehr 1956; Wang 1986 gives a vivid firsthand account of the difficulties involved in acquiring Jannings's collection of bronze vessels and weapons.

58. Kelley and Ch'en 1946. For additional details about the collection and its donor, see McCormick and Smith 1939; Kelley 1939; "Kate S. Buckingham" 1945.

59. Lodge, et al. 1946.

60. The bronzes purchased by Charles Lang Freer were not published until 1967. See Pope, et al. 1967.

61. Lodge, et al. 1946, 40.

62. Karlgren 1952. For information about the Pillsbury collection and a review of the catalogue, see *Bulletin of the Minneapolis Institute of Arts* 1950:39(18):86–91; *Bulletin of the Minneapolis Institute of Arts* 1952:41(24):126–36; Loehr 1953b.

63. van Heusden 1952.

64. See Liang 1944.

65. Kidder 1956. Perhaps the most controversial of all American catalogues on Chinese bronzes, this volume provoked a series of articles. See Loehr 1958; *Artibus Asiae* 1958:21(1):91–94; Kidder 1958; Grossman 1958; Davidson 1961.

66. Another example of this type of *gui* is in the British Museum. See Rawson 1987, 78–79.

67. "A Bronze Sacrificial Vessel" 1916.

68. Duanfang 1908.

69. Bagley 1987.

70. Umehara 1947. In his introduction to the catalogue, Umehara Sueji, who apparently never saw any of the original bronzes, singles out the Xiaochen Yi *jia*, noting that its inscription makes it the most important piece in the collection.

71. Shang 1935, *zhu* 11b–12b.

72. Bagley 1987, 293–309.

73. Kidder 1956, 38.

74. Loehr 1968, 106.

75. Cheng, et al. 1989, figs. 15–17. One major difference between the Saint Louis *zhi* and the one in China is the two-character inscription—*zi fu*—on the bottom of the interior. For a list of other vessels on which this same inscription appears, see Chen Mengjia 1977, 1:120–21, A639. All the inscriptions cited by Chen Mengjia were reproduced in Chinese compendia readily available to forgers.

76. Chen Mengjia 1950.

77. Apparently Chen Mengjia was referring to the two gilt bronze *hu* in the Buckingham Collection acquired in 1931 (pls. 27–30) in the catalogue prepared by Chen and Charles Kelley (Kelley and Ch'en 1946), where they are designated as "probably Han dynasty," and the *hu* in the Museum of Fine Arts, Boston, purchased in 1929 (Tomita 1930; the bronze is illustrated on the cover). A fourth gilt bronze is in the Study Collection of the Freer Gallery of Art. Cheung Kwong-yue refers to one of the gilt bronze *hu* in the Buckingham collection as a forgery made in Suzhou by Zhou Meiku. See Cheung 1974, 72.

78. See Ferguson 1928; and "Bronze Vessels" 1929, a review of Yetts 1929.

79. Chase 1991, 36.

80. Loehr 1953a.

81. Loehr 1968.

82. Wen Fong 1980. Contributors to the catalogue were Ma Chengyuan, Kwang-chih Chang, Robert L. Thorp, Robert W. Bagley, Jenny F. So, and Maxwell K. Hearn.

83. There is some irony in the fact that the Metropolitan should have been such a catalyst, especially since the initial venture of the museum in collecting Chinese bronzes was so tentative. In 1912, authorities at the Metropolitan asked John C. Ferguson, a longtime resident of China and a well-known authority on Chinese art, to secure "representative specimens of Chinese art." In response to that request, Ferguson purchased several bronzes for the museum. When the Trustees saw those bronzes, however, they raised some questions about whether they might be better housed in an archaeological museum rather than in a museum of fine arts. Fortunately no drastic reversal of policy resulted from that cool response and the Metropolitan Museum today houses a fine group of Chinese bronzes. See Ferguson 1928; and "Bronze Vessels" 1929.

84. Delbanco 1983.

85. Bagley 1987. For reviews of Bagley's book see Lawton 1988; and Childs-Johnson 1989.

86. Rawson 1990. For reviews of Rawson's book see Mackenzie 1991; and von Falkenhausen 1993.

87. So 1995.

88. "Guanghan Sanxingdui" 1987; Shen 1987; Wen Jiao 1987; Bagley 1988; Bagley 1990; Chen Xiandan 1990.

89. Han 1991; Peng, et al. 1991; Bagley 1992.

References

Ausstellung chinesischer Kunst 1929
Ausstellung chinesischer Kunst veranstaltet von der Gesellschaft für Ostasiatische Kunst und der Preussischen Akademie der Künste Berlin. 1929. Berlin: Würfel Verlag.

Ayscough 1929
Ayscough, Francis. 1929. "Phantasmagoria." *Ostasiatische Zeitschrift*, N.F.5:46–48.

Bagley 1987
Bagley, Robert W. 1987. *Shang Ritual Bronzes in the Arthur M. Sackler Collections*. Ancient Chinese Bronzes in the Arthur M. Sackler Collections, vol. 1. Cambridge: Harvard University Press.

Bagley 1988
——. 1988. "Sacrificial Pits of the Shang Period at Sanxingdui in Guanghan County, Sichuan." *Arts Asiatiques* 43:78–86.

Bagley 1990
——. 1990. "A Shang City in Sichuan Province." *Orientations* 21(11):52–67.

Bagley 1992
——. 1992. "Changjiang Bronzes and Shang Archaeology." In *International Colloquium on Chinese Art History, 1991: Proceedings, Antiquities, Part 1*, 209–55. Taiwan: National Palace Museum.

Bobot 1983
Bobot, Marie-Thérèse. 1983. *Musée Cernuschi: Promenade dans les collections chinoises*. Paris: Musée Cernuschi.

"A Bronze Sacrificial Vessel" 1916
"A Bronze Sacrificial Vessel of the Chou Dynasty in China." 1916. *Bulletin of the City Art Museum of St. Louis* 2(3):9.

"Bronze Vessels" 1929
"Bronze Vessels." 1929. *The Chinese Journal of Science and Arts* 11(6):284–88.

Catalogue of an Exhibition 1916
Catalogue of an Exhibition of Early Chinese Pottery and Sculpture. 1916. New York: The Metropolitan Museum of Art.

"Catalogue of the Exhibition" 1953
"Catalogue of the Exhibition of Early Chinese Bronzes." 1953. *Transactions of the Oriental Ceramic Society* 26.

Chase 1991
Chase, W. Thomas. 1991. *Ancient Chinese Bronze Art: Casting the Precious Sacral Vessel*. New York: China House Gallery.

Chen Mengjia 1950
Chen Mengjia. 1950. "Zhongguo gudai tongqi senyang dao Meiguo qude?" *Wenwu cankao ziliao* (11):87–91

Chen Mengjia 1977
——, ed. 1977. *In Shu seidōki bunrui zuroku*. 2 vols. Tokyo: Kyūko Shorin.

Chen Xiandan 1990
——. 1990. "Guanghan Sanxingdui qingtongqi yanjiu." *Sichuan wenwu* (6):22–30.

Cheng, et al. 1989
Cheng Changxin, Wang Yongchang, and Cheng Ruixiu. 1989. "Tongqi bian wei qianshuo (zhong)." *Wenwu* (11):33–34.

Cheung 1974
Cheung Kwong-yue. 1974. *Weizuo xian Qin yiqi mingwen shuyao*. Hong Kong: Xianggang Shuju.

Childs-Johnson 1989
Childs-Johnson, Elizabeth. 1989. *The Art Bulletin* 71(1):149–56.

Chinese Bronzes 1938
Chinese Bronzes of the Shang (1766–1122 B.C.) Through the T'ang Dynasty (A.D. 618–906): An Exhibition Lent by American Collectors and Museums. 1938. New York: The Metropolitan Museum of Art.

Collection Hayashi 1903
Collection Hayashi: Objets d'Art. Part 2. 1903. Paris: Hôtel Drouot.

Davidson 1961
Davidson, J. LeRoy. 1961. *Ars Orientalis* 4:430–32.

Delbanco 1983
Delbanco, Dawn Ho. 1983. *Art from Ritual: Ancient Chinese Bronze Vessels from the Arthur M. Sackler Collections*. Cambridge: The Fogg Art Museum and The Arthur M. Sackler Foundation.

Duanfang 1908
Duanfang. 1908. *Taozhai jijin lu*. Shanghai: Yuzheng Shuju.

Duret 1874
Duret, Théodore. 1874. *Voyages en Asie: Le Japon—La Chine—La Mongolie—Java—Ceylan—L'Inde*. Paris: M. Levy.

Ecke 1939
Ecke, Gustav. 1939. *Frühe chinesische Bronzen aus der Sammlung Oskar Trautmann*. Beijing: Furen University.

Elisseeff 1977   Elisseeff, Vadime. 1977. *Bronzes archaïques chinois au Musée Cernuschi*. Paris: Musée Cernuschi.

Erkes 1940   Erkes, Eduard. 1940. *Artibus Asiae* 8(1):281–83.

Eugene and Agnes E. Meyer 1971   *Eugene and Agnes E. Meyer Memorial Exhibition*. 1971. Washington, D.C.: Freer Gallery of Art.

Exhibition of Chinese Art 1925   *Exhibition of Chinese Art: Amsterdam 1925, Municipal Museum, 13 September–18 October*. 1925. The Hague: N.V. Drukkerij Ten Hagen.

"Exhibition of Early Chinese Bronzes" 1934   "Exhibition of Early Chinese Bronzes." 1934. *Bulletin of the Museum of Far Eastern Antiquities, Stockholm* 6:81–136.

Ferguson 1928   Ferguson, John C. 1928. "The Four Bronze Vessels of the Marquis of Ch'i." *Eastern Art* 1(1):43–47.

Fischer 1935   Fischer, Otto. 1935. *Artibus Asiae* 4(1):314–15.

Grossman 1958   Grossman, Betty. 1958. "*Early Chinese Bronzes* by J. Edward Kidder, Jr." *Bulletin of the City Art Museum of St. Louis* 43(1):6–9.

Grousset 1937   Grousset, René. 1937. *L'Évolution des bronzes chinois archaïques*. Paris: Les Éditions d'art et d'histoire.

"Guanghan Sanxingdui" 1987   "Guanghan Sanxingdui yizhi yihao jisi keng fajue jianbao." 1987. *Wenwu* (10):1–15.

Hakkakujō 1907   *Hakkakujō*. 1907. 2 vols. Kyoto: P.p.

Han 1991   Han Kangxin. 1991. "Jiangxi Xin'gan Dayangzhou Shang mu fajue jianbao." *Wenwu* (10):1–23.

Hayashi 1984   Hayashi Minao. 1984. *In Shū jidai seidōki no kenkyū: In Shū seidōki sōran 1*. 2 vols. Tokyo: Yoshikawa Kōbunkan.

Hentze 1927   Hentze, Carl. 1927. *Artibus Asiae* 2(1):66–69.

Ichikawa 1848   Ichikawa Beian. 1848. *Shōsanrindō shoga bunbō zuroku*. 10 vols. Kobe: P.p.

Illustrated Catalogue 1913   *Illustrated Catalogue of the Remarkable Collection of the Imperial Prince Kung of China: A Wonderful Treasury of Celestial Art Recently Acquired by the Widely Known Firm of Yamanaka & Company*. 1913. New York: The American Art Association.

Illustrated Catalogue 1914   *Illustrated Catalogue of the Remarkable Collection of Ancient Chinese Bronzes, Beautiful Old Porcelains, Amber and Stone Carvings, Sumptuous Eighteenth Century Brocades, Interesting Old Paintings on Glass, and Fine Old Carpets, Rugs and Furniture, from Ancient Palaces and Temples of China*. 1914. New York: The American Art Association.

International Exhibition 1935–36   *International Exhibition of Chinese Art: Catalogue and Illustrated Supplement*. 1935–36. London: Royal Academy of Arts.

Karlgren 1936   Karlgren, Bernhard. 1936. "Yin and Chou in Chinese Bronzes." *Bulletin of the Museum of Far Eastern Antiquities, Stockholm* (8):9–154.

Karlgren 1952   ———. 1952. *A Catalogue of the Chinese Bronzes in the Alfred F. Pillsbury Collection*. Minneapolis: Minneapolis Institute of Arts.

"Kate S. Buckingham" 1945   "Kate S. Buckingham As a Collector." 1945. *Bulletin of the Art Institute of Chicago* 39(1):2–5.

Kelley 1939   Kelley, Charles Fabens 1939. "The Buckingham Chinese Collection." *Bulletin of the Art Institute of Chicago* 33(4):52–54.

Kelley and Ch'en 1946   Kelley, Charles Fabens, and Ch'en Meng-chia (Chen Mengjia). 1946. *Chinese Bronzes from the Buckingham Collection*. Chicago: The Art Institute of Chicago.

Kerr 1990   Kerr, Rose. 1990. *Later Chinese Bronzes*. London: Victoria and Albert Museum.

Kidder 1956   Kidder, J. Edward, Jr. 1956. *Early Chinese Bronzes in the City Art Museum of St. Louis*. St. Louis: City Art Museum.

Kidder 1958   ———. 1958. "Rejoinder." *Artibus Asiae* 21(1):94–95.

Koop 1924 Koop, Albert J. 1924. *Early Chinese Bronzes*. London: Ernest Benn Ltd.

Kümmel 1928 Kümmel, Otto. 1928. *Chinesische Bronzen aus der Abteilung für Ostasiatische Kunst aus den Staatlichen Museen Berlin*. Berlin: Graphische Kunstanstalt Albert Frisch.

Kümmel 1930a ————. 1930. *Chinesische Kunst*. Berlin: Bruno Cassier Verlag.

Kümmel 1930b ————. 1930. *Jörg Trübner zum Gedächtnis: Ergebnisse seiner letzten chinesischen Reisen*. Berlin: Klinkhardt and Biermann.

Kümmel 1941 ————. 1941. *Ostasiatische Zeitschrift*, N.F.17(1–2):66.

Kuwabara 1967 Kuwabara Sumio. 1967. "Seikai ichi no Tōyō kobijutsu-shō: Yamanaka Shōkai seisuiki." *Geijutsu Shinchō* 18(1):153–62.

Lawton 1988 Lawton, Thomas. 1988. *Orientations* 19(10):73–77.

Lawton 1995 ————. 1995. "Yamanaka Sadajirō: Advocate for Asian Art." *Orientations* 26(1):80–93.

Lawton 1997 ————. 1997. *A Time of Transition: Two Collectors of Chinese Art*. Lawrence: The Spencer Museum of Art, University of Kansas.

Lehnert 1909 Lehnert, Georg Hermann, ed. 1909. *Illustrierte Geschichte des Kunstgewerbes*. 2 vols. Berlin: Verlag von Martin Oldenbourg.

Liang 1944 Liang Shangchun. 1944. *Yanku jijin tulu*. Beijing: N.p.

Lippe 1950 Lippe, Aschwin. 1950. "A Gift of Chinese Bronzes." *The Metropolitan Museum of Art Bulletin* 9(4):97–107.

Liu 1989 Liu Ling. 1989. *Qianlong sijian zonglibiao*. Beijing: Zonghua Shuju.

Lodge, et al. 1946 Lodge, John E., Archibald G. Wenley, and John A. Pope. 1946. *A Descriptive and Illustrative Catalogue of Chinese Bronzes Acquired During the Administration of John Ellerton Lodge*. Washington, D.C.: Freer Gallery of Art.

Loehr 1936 Loehr, Max. 1936. "Beiträge zur Chronologie der älteren chinesischen Bronzen." *Ostasiatische Zeitschrift* N.F.12:3–41.

Loehr 1948 ————. 1948. *Monumenta Serica* 13:420–26.

Loehr 1953a ————. 1953. "The Bronze Styles of the Anyang Period (1300–1028 B.C.)." *Archives of the Chinese Art Society of America* 7:42–53.

Loehr 1953b ————. 1953. *The Far Eastern Quarterly* 12(2):332–35.

Loehr 1956 ————. 1956. *Chinese Bronze Age Weapons: The Werner Jannings Collection in the Chinese National Palace Museum, Peking*. Ann Arbor: The University of Michigan Press.

Loehr 1958 ————. 1958. *Journal of Asian Studies* 17(2):281–83.

Loehr 1968 ————. 1968. *Ritual Vessels of Bronze Age China*. New York: Asia Society.

Loo 1940 Loo, C.T. 1940. Preface. *An Exhibition of Chinese Stone Sculpture*. New York: C.T. Loo.

Luo Tan (Lawton) 1992 Luo Tan (Thomas Lawton). 1992. "Yizu Zhongguo yishu shoucangpin de fen yu he." *Shanghai Bowuguan jikan* 6(October):421–34.

Mackenzie 1991 Mackenzie, Colin. 1991. *Ars Orientalis* 21:153–54

McCormick and Smith 1939 McCormick, Chauncey, and Walter B. Smith. 1939. "Kate Sturges Buckingham." *Bulletin of the Art Institute of Chicago* 33(4):50– 51.

Menzies 1940 Menzies, James M. 1940. "The Appreciation of Chinese Bronzes." In *An Exhibition of Ancient Chinese Ritual Bronzes*. New York: C.T. Loo.

Mizuno 1968 Mizuno Seiichi. 1968. *Tōyō bijutsu 5: Dōki*. Tokyo: Asahi Shinbunsha.

Morgan 1929	Morgan, Evan. 1929. *Journal of the North China Branch of the Royal Asiatic Society* 60:129–35.
Münsterberg 1924	Münsterberg, Oskar. 1924. *Chinesische Kunstgeschichte*. 2 vols. Esslingen: Paul Neff Verlag.
Niu and Wang 1989	Niu Deming and Wang Xuezhen. 1989. "Gugong guibao guangzhao yinglun—yi 1935 nian 'Lundon Zhongguo yishu guoji zhanlanghui.'" *Wenwu tiandi* (6):23–27.
Okada 1953	Okada Yoshisaburō. 1953. "Hōkei no shoki ni tsuite: Chūgoku kodōku shūsei no hitotsu no kokoromi." *Tōhō gakuhō* 23:219– 30.
Pelliot 1930	Pelliot, Paul. 1930. *T'oung pao* 2(27):359–406.
Peng, et al. 1991	Peng Shifan, Liu Lin, and Zhan Kaixun. 1991. "Guanyu Xin'gan Dayangzhou Shang mu niandai wenti de tantao." *Wenwu* 10:24–32.
Pope, et al. 1967	Pope, John Alexander, et al. *The Freer Chinese Bronzes*. 2 vols. Washington, D.C.: Freer Gallery of Art.
Priest 1938	Priest, Alan. 1938. "An Exhibition of Chinese Bronzes from American Collections." *Bulletin of The Metropolitan Museum of Art* 33(10):216–22.
Rawson 1987	Rawson, Jessica. 1987. *Chinese Bronzes: Art and Ritual*. London: The British Museum.
Rawson 1990	———. 1990. *Western Zhou Ritual Bronzes from the Arthur M. Sackler Collections*. Ancient Chinese Bronzes in the Arthur M. Sackler Collections, vols. 2A–B. Cambridge: Harvard University Press.
Reitz 1924	Reitz, S.C. Bosch. 1924. "The Tuang Fang Sacrificial Table." *Bulletin of The Metropolitan Museum of Art* 19(6):141–44.
Rong 1935	Rong Geng. 1935. *Haiwai jijin tulu*. Beijing: Yanjing Daxu Kaogu Xueshe.
Rong 1941	———. 1941. *Shang Zhou yiqi tongkao*. 2 vols. Beijing: Harvard-Yenching Institute.
Salles 1934a	Salles, Georges. 1934. *Bronzes chinois des dynasties Tcheou, Ts'in, et Han*. Paris: Musée de l'Orangerie.
Salles 1934b	———. 1934. "Les bronzes de Li-yü." *Revue des arts asiatiques* 8(3):146–58.
Sammlung Lochow 1943	*Sammlung Lochow: Chinesische Bronzen I*. 1943. Ed. Gustav Ecke. Beijing: Furen University.
Sammlung Lochow 1944	*Sammlung Lochow: Chinesische Bronzen II*. Ed. Hans Jürgen von Lochow. Beijing: Furen University.
Segi 1991	Segi Shin'ichi. 1991. "Bijutsu māketto sōseiki." *Geijutsu Shinchō* 42(12):4–8.
Sen'oku seishō 1911–16	*Sen'oku seishō*. 1911–16. Kyoto: P.p.
Shang 1935	Shang Chengzuo. 1935. *Shierjia jijin lu*. Nanjing.
Shen 1987	Shen Zhongchang. 1987. "Sanxindui Erhao jisi keng qingtong li renxiang chuji." *Wenwu* (10):16–17
Sirén 1933	Sirén, Osvald. 1993. *Bulletin of the School of Oriental Studies* 7, pt. 1:192–203.
So 1995	So, Jenny F. 1995. *Eastern Zhou Ritual Bronzes from the Arthur M. Sackler Collections*. Ancient Chinese Bronzes in the Arthur M. Sackler Collections, vol. 3. New York: Abrams.
Sumitomo 1955	Sumitomo Kichizaemon. 1955. *Sumitomo Shunsui*. Kyoto: Sumitomo Shunsui Hensan Iinkai.
Tch'ou 1924	Tch'ou Tö-yi (Chu Deyi). 1924. *Bronzes antiques de la Chine appartenant à C.T. Loo et Cie*. Paris: G. van Oest.
Tomita 1930	Tomita, Kojiro. 1930. "A Bronze Jar of the Early Han Period." *Bulletin of the Museum of Fine Arts, Boston* 28(167):46–7.
Trübner 1929	Trübner, Jörg. 1929. *Yu und Kuang: Zum Typologie der chinesischen Bronzen*. Leipzig: Klinkhardt und Biermann.

Umehara 1933   Umehara Sueji. 1933. *Henkin no kōkogaku-teki kōsatsu/Étude archaeologique sur le Pien-chin, ou série de bronzes avec une table pour l'usage rituel dans la Chine antique*. Memoire de Tōhō-Bunka-Gakuin Kyōto Kenkyūsho, vol 2. Kyoto: Tōhō Bunka Gakuin Kyōto Kenkyūsho.

Umehara 1947   ———. 1947. *Kankarō kikkinzu*. Kyoto: P.p.

Umehara 1959   ———. 1959. "Sensei-shō Hōkei-ken shutsudo no dai ni no henkin." *Tōhōgaku kiyō* 1:1–15.

van Heusden 1952   van Heusden, Willem. 1952. *Ancient Chinese Bronzes of the Shang and Chou Dynasties*. Tokyo: P.p.

Visser 1926   Visser, H.F.E., ed. 1926. *The Exhibition of Chinese Art of the Society of Friends of Asiatic Art, Amsterdam 1925*. The Hague: Martinus Nijhoff.

von Falkenhausen 1993   von Falkenhausen, Lothar. 1993. *Early China* 18:139–226.

Wang 1986   Wang Shixiang. 1986. "Huiyi kangzhan shengli hou wo can yu de wenwu qingli gongzuo (2)." *Wenwu tiandi* (6):28–31.

Weisberg 1986   Weisberg, Gabriel P. 1986. *Art Nouveau Bing: Paris Style, 1900*. New York: Abrams.

Wen Fong 1980   Wen Fong, ed. 1980. *The Great Bronze Age of China*. New York: The Metropolitan Museum of Art.

Wen Jiao 1987   Wen Jiao. 1987. "Ancient Bronze Statues Discovered in Sichuan." *China Reconstructs* (6):20–23.

Weng and Yang 1982   Weng Wange and Yang Boda. 1982. *The Palace Museum*. New York: Abrams.

Yamanaka 1939   *Yamanaka Sadajirō-den*. 1939. Osaka: Ko Yamanaka Sadajirō-den Hensankai.

Yetts 1929   Yetts, W. Perceval. 1929. *The George Eumorfopoulos Collection: Catalogue of the Chinese and Corean Bronzes, Sculpture, Jade and Jewelry, and Miscellaneous Objects, Vol. 1: Bronzes: Ritual and Other Vessels, Weapons, Etc.* London: Ernest Benn Ltd.

Yetts 1939   ———. 1939. *The Cull Chinese Bronzes*. London: University of London, Courtauld Institute of Art.

No. 1

Gu

late 14th–early 13th century BC
Shang dynasty
Inscribed: *X*
Height: 21.7 cm
Mouth diameter: 12.5 cm
Gift of J. Lionberger Davis
29:1951

The compacted tripart design of this bronze wine beaker or *gu* has long been considered its most distinctive feature, setting it apart from other *gu* as a rare example of an elegant but common ancient Chinese ritual vessel.[1]

The *gu* vessel shape generally is circular in cross section and is notable for its flared mouth and foot. In the Saint Louis piece, the vessel's mouth is wide and generous; its slight rim is created by a sharp vertical bevel. The ample dimensions of the mouth are sustained in the body by the gradual downward curve of the rim to the columnar waist, at which the curve begins its outward splay to the foot. The full and complementary proportions of the *gu* provide a solid and robust character to the vessel shape; the outline is broken only by the low relief of the cast surface designs.

The designs are created by relief lines in tight curvilinear patterns of hooks, volutes, and curls arranged to form creatures with staring eyes. The lower half of the *gu* is marked by a single raised band that encircles the vessel at midpoint. Below are three successive registers which follow one another without interruption. The waist hosts a frieze of four zoomorphic motifs bounded top and bottom by a row of small circlets. Divided into pairs, the animals confront one another, separated only by a low narrow flange. The joined features of the two beasts form a *taotie* with wide horns and forepaws. The second register is comprised of two large zoomorphs seen with the head of one animal to the tail of the other, both with large round eyes in low relief. Lastly, four smaller but similar animal move head-to-tail around the foot.

The vessel was cast in a two-part mold and is perforated at the second design register by a pair of cast four-lobed ovals longitudinally aligned and arranged opposite one another. The surface patina is generally smooth with encrustations of malachite and cuprite. Tan-brown and olive areas suggest that the rim has been restored; some areas of the low-relief design also appear to have undergone restoration.

The inscription is in the form of a single animal, shaped as a slight depression with a raised edge, inside the foot. The creature is seen in profile and bears an elongated snout and lower jaw, a three-toed hind foot, and a long looping tail. Symbols of this nature are usually considered clan signs.

The decor is rendered in an early form of Loehr Style III. Its rare compacted animal forms with no breaks between registers makes it distinct from the designs of both earlier and later *gu*. However, two other *gu* bearing undivided tripart designs are known: the late Style II or early Style III vessel from Shaanxi Zichang with similar but slightly less distinct lip and foot rims; and a bronze from Panlongcheng.[2]

The Saint Louis *gu* dates from circa early 13th or late 14th century BC. Its full broad shape relates directly to similar early *gu* excavated from Taixicun, Gaocheng, Hebei, a site which represents a transitional phase between the Erligang and Anyang.[3] The common features of these *gu* include the ample, somewhat stocky proportions; the narrow vertical lip rim; and the corresponding narrow rise at the foot rim, as well as the four-lobed apertures piercing the vessel below the waist.

Selected Bibliography
Kidder 1956, 31–32, pl. 1; Loehr 1968, 42–43, cat. no. 13; *Handbook* 1975, 268; Barnard and Cheung 1978, 882; Buchanan and Ronan 1980, 149; *Haiwai* 1985, 48, pl. 48.

Notes
1. Max Loehr was the first to note the special character of this *gu* (Loehr 1968, 42–43, cat. no. 13).

2. *Shaanxi chutu* 1979, no. 4; and *Wenwu* 1976(2):31, figs. 30.7, 30.9 respectively.

3. Rawson 1987, 17.

No. 2

Jia

13th century BC
Shang dynasty
Inscribed: *Jiu Yi*
Height: 33.9 cm
Mouth diameter: 23.3 cm
Gift of J. Lionberger Davis
224:1950

Supported by three thickly cast legs, the short compressed body of this *jia* constricts slightly and flares upward to form a deep wide mouth. From opposite sides of the rim, two heavy square posts rise flush along the molded mouth rim, both capped by large nippled bosses which are further decorated by a pattern of swirling hooks. The flat strap handle is slightly rounded with a thin mold mark running along its length. The tapered pointed legs are lozenge-shaped in cross section and undecorated.

The two registers of design on the *jia* are flush with the surface of the vessel and symmetrically arranged on the same axes, the main ornamentation of three *taotie* being aligned directly below a narrower band of similar but smaller motifs. In the larger register on the body, the plain round boss eyes of the large zoomorphs are divided by a low, arching, flat ridge; the eyes are set in hooked canthi below recumbent T-curls that delineate the horns. The short straight back of the animal is decorated with a set of vertical quills that introduce the curled extended tail. A row of sharp, hooking, needlelike teeth mark the curved upper jaw. Flanking the zoomorphs, the heads and bodies of subsidiary animals are suggested by small round bosses near the bottom of the band.

The design of the upper register is similar to that of the lower band; its components are slightly reduced in scale and have a greater diagonal orientation, but on the whole it is directly related to the design on the body below. A notable difference in the upper ornamental band is the absence of teeth in the jaws of the confronting animals and the oval instead of round shapes of the small secondary eyes.

The patina of the bronze is a mottled dark leaden-green color that is encrusted with a thin layer of malachite and cuprite. The interstices and surfaces are filled with a light-beige colored earth which darkens in patches on the decorated registers.

The two-character inscription is cast in intaglio in a single column on the wall of the body just beneath the loop of the closed handle. The inscription reads: *Jiu Yi*, "libations for [Father or Ancestor] *Yi*."[1]

The Saint Louis *jia* is rendered in a well-developed Loehr Style III. This style is characterized by a plethora of fine, even minute, figural details in the form of quills, volutes, and bosses, all of which are symmetrically disposed and intimately descriptive of the image's zoomorphic attributes.[2] The fanged mask of the main register is related to an earlier design introduced by pre-Anyang bronzes and carried on into the Anyang phase in ivory, horn, and bone carving.[3] Other features of the early design include triangular decorative panels below the fangs, parallel ribbons, plain shieldlike devices at the flanges, single-band bodies with spurred curls, tightly formed claws, and T-curls.[4]

A highly unusual and unexpected feature of the Saint Louis *jia* is the asymmetry of four of its six masks. Ordinarily, on Shang and Zhou bronzes the *taotie* is created by two confronting but identically poised animals in profile. This is not the case with the Saint Louis *jia*, which conforms to a different but nonetheless perfectly balanced design scheme.

On the Museum's bronze, the two masks depicted on the face opposite the strap handle are symmetrical in that the tails of the four confronting animals curl outward beyond their bodies. The four asymmetrical masks are aligned near the handle and separated by the inscription. Each of these is composed of two zoomorphic profiles: one animal with the tail curled over its back and the other with the tail curling outward. Balance of design is achieved by the joining of all adjacent animals with outward curling tails at their tails. The remaining animals with tails curling over their backs are all positioned with tails aligned at the inscription and handle.[5]

This *jia* is an accomplished and innovative rendering of the earlier toothed mask design in Loehr Style III and most likely dates from the early Anyang phase, through Yinxu period I and into period II.

Selected Bibliography

Luo 1937, 13:49a; Hoopes 1951, fig. 3; *Handbook* 1953, 250; Kidder 1956, 54–55, pl. 12; Loehr 1968, 44–45, cat. no. 14; Barnard and Cheung 1978, 873, no. 1505.

Notes

1. The character *jiu* means "fermented liquor," which in the ritual context and dedication of a vessel to an ancestor would be a sacrificial libation (Loehr 1968, 44, cat. no. 14; see also Kidder 1956, 54). There are three comparable inscriptions in Barnard and Cheung 1978, 873, nos. 1602–4. An apparently engraved version of the Saint Louis inscription occurs on a *jia* (M.958.374) in the Hood Museum of Art, Dartmouth College (by formal correspondence, 22 May 1981).

2. As in Styles I and II, there is in Loehr Style III an absence of extraneous or superfluous line. There appears to be no discernible ground, as every motif appears to have corporeal significance either as a body part or embellishment to a body part. Only with the development of Style IV is there a differentiation between image and ground. A Loehr Style III *jia* (R2038) of very comparable shape, size (24 cm), design, and proportion was excavated from tomb M232 at Anyang (Li and Wan 1968, pl. 3).

3. Compare the well-known carvings on ivory, horn, and bone from tomb HPKM1001 at Houjiazhuang, Anyang (Gao Quxun 1962, 1:269–passim and 2: pls. 202–6).

4. For a detailed discussion of this distinctive early group of Loehr Style IV, see Huber 1981b, especially 19–27.

5. For another example of this unusual phenomenon compare the asymmetrical *taotie* on a *jue* in *Kaogu xuebao* 1979(1):27–146.

No. 3

Gu

13th century BC
Shang dynasty
Inscribed: *Niao*
Height: 24.4 cm
Mouth diameter: 14.5 cm
Gift of J. Lionberger Davis
28:1951

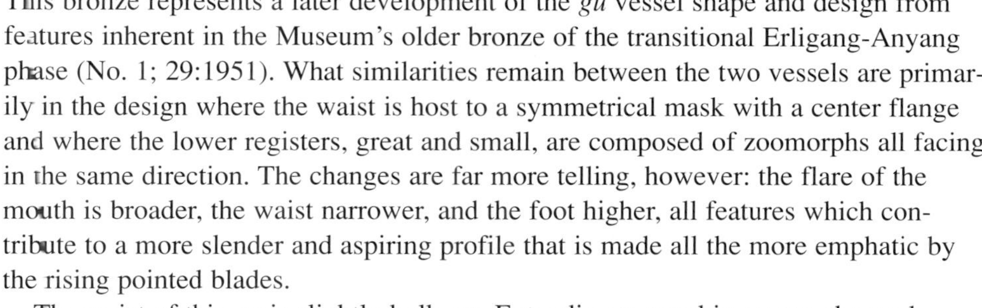

This bronze represents a later development of the *gu* vessel shape and design from features inherent in the Museum's older bronze of the transitional Erligang-Anyang phase (No. 1; 29:1951). What similarities remain between the two vessels are primarily in the design where the waist is host to a symmetrical mask with a center flange and where the lower registers, great and small, are composed of zoomorphs all facing in the same direction. The changes are far more telling, however: the flare of the mouth is broader, the waist narrower, and the foot higher, all features which contribute to a more slender and aspiring profile that is made all the more emphatic by the rising pointed blades.

The waist of this *gu* is slightly bulbous. Extending upward in an arc, the neck forms at length into a trumpet mouth. Below, the foot flares outward and ends in a molded foot rim. Four decorative blades rise from the neck to the underside of the mouth rim. The blades, which start from a narrow band of connected and spiraling C-curls, are composed of inverted V-shapes with hooking, tightly wound volutes on the interiors and sharp spurs on the exteriors; near the tip of the blades are plain points. The designs are rendered in thin ribbons against a ground of irregular *leiwen*. The blades are separated from the waist by a plain band. Four short scored flanges divide the waist into equal sections, each decorated with the profile of a single bird in low relief that fills the space. These birds are paired on the vessel as confronting animals with small curling tufts on their heads; the pairs are separated by a plain shield-like device set on either side of a flange. The eyes of the fowl are small round bosses below which are half-moon shapes on the neck and smaller half-moons on the wings and bodies; the beak is marked by two chevrons. The breast of the bird is filled with a large spurred volute. A wide plain band, which is decorated with two bowstrings, is interrupted by an open cruciform on one side and an irregular hole on the other. As with the waist, the foot is divided by neatly scored flanges into four parts. The top portion of each section bears a small horned dragon in profile with a sharply hooked beak, a large curling tail over its back, and a clawed forefoot. These four small dragons form an ornamental band of animals facing in the same direction; the band is separated from the larger register by a rather fine ribbon. The four large dragons in profile face the same direction as the smaller zoomorphs. Their eyes are plain round bosses set in large outer canthi from which sprout backward-curving horns and large tines. Curling probocides form their snouts and jaws, while from their truncated bodies extend forelegs and claws. The zoomorphs are set against a varied *leiwen* ground that is flush with the surfaces of the primary designs.

The mottled gray-green patina is overlaid with patches of fine granular malachite and azurite. There are occasional spots of cuprite and beige-colored earth.

The vessel is inscribed with a single pictograph of a bird in intaglio on the inside of the foot just above the footring. The design of the bird follows rather closely the rendering of the fowl on the vessel with its pair of chevrons on the sharp curved beak, the curling tuft at the back of the head, and distinctly split body and wing. The modern reading of the graph is *niao*, "bird."

Unlike the profusion of linear form and visual integration of Loehr Style III that is represented by the older vessel (No. 1), this *gu* is an example of Loehr Style IV, a style in which large motifs clearly stand out against a ground of smaller patterns, all flush to the surface of the vessel. However, the design of this bronze is nonetheless notable for its low-relief birds which anticipate the later and more dramatic relief designs of Loehr Style V.

The unusual placement of these blades is noteworthy. Although uncommon, it is not unknown for the blades themselves to be aligned off center of the flanges, their daggerlike shapes extending without interruption from each of the lower decorated segments to the top of the vessel; thus, the V-points formed between the blades, rather than the blades themselves, are centered on the flanges.

Regarding its distinctive design, the Saint Louis bronze is part of a small but singular group of works dating from the early Anyang phase that are related by the selection of particular motifs and the special manner in which the elements are rendered.[1] The *taotie* and dragons on these bronzes have plain round bosses for eyes and extended vertical or horizontal bodies composed of a single band ornamented with spurred T-curls resembling the letter C intersected by a stem. The taut quality of the design is directly related to the manner of depiction, in which the animals have tightly curled, almost concentrically disposed, probocides and tails, and claws and dewclaws which are sharply delineated and form nearly closed circles. The larger features of the zoomorphs are often created with triple parallel lines or ribbons, while the *leiwen* are irregular voluted filler, sometimes taking the form of elongated geometrical frets. The group is further marked by rising blades which have multiple exterior hooks and spurs and small plain shieldlike figures near the tops of flanges. There is a paucity of excavated material with this exceptional design;[2] most of the known examples are of unknown provenance and in public and private collections. But on the basis of style, the Saint Louis *gu* is likely a transitional piece dating to the very beginning of the Anyang phase.[3]

Selected Bibliography
Hoopes 1951, figs. 1–2; Kidder 1956, 30, pl. 1.

Notes
1. For a detailed discussion of this distinctive early group of Loehr Style IV bronzes see Huber 1981b, especially 19–27.

2. The design is exemplified among archaeological finds by an early Anyang-phase *hu* (C7M9:1) vessel from the 1954 excavation at People's Park or Renmin gongyuan at Zhengzhou (*Henan chutu* 1981, 56, cat. no. 345; 273, pl. 345). Later manifestations of the design include the *ding* excavated from a subsidiary grave (E14) in the eastern section of the large tomb (WKGM1) at Wuguancun (*Henan chutu* 1981, 45, cat. no. 276; 221, pl. 276). A closely related but different style of design can be found in Shang pottery and ivories of the Anyang phase (see Gao Quxun 1967, pls. 202.2, 203.2, 203.10).

3. Huber dates the *gu* in the Royal Ontario Museum and the *ding* excavated from Wuguancun to the earliest Yinxu period during the reigns of the first Anyang ruler Pan Geng or his brothers Xiao Xin or Xiao Yi (Huber 1981b, 25). Stylistically, both the *gu* and *ding* are later manifestations of this particular design. If Huber's dating of the two bronzes is correct, it would therefore indicate a pre-Anyang date for all earlier examples of the design including the Saint Louis *gu*.

Gu No. 3 (28:1951)

No. 4

Jiao

13th century BC
Shang dynasty
Inscribed: *Tian Ding*
Height: 24.4 cm
Mouth diameter: 16.7 cm
Gift of J. Lionberger Davis
220:1950

Widely agape, the mouth of this *jiao* features distinctive double points unequal in size, one being slightly larger than the other. From either point, the shape of the vessel narrows to an ovoid body which is elliptical in cross section and supported by three thick heavy legs of bladelike pointed form. The wine container has a handle, the broad upper portion of which is surmounted by an animal head in low relief. The sculpted horns of the head are convoluted and formed by parallel thin ribbons that end in raised tips. The animal's eyes and sawteeth are simply rendered in intaglio and flanked by dimpled pointed ears.

The *jiao* is ornamented by two registers of decor on the body proper and below the mouth rim. The primary designs are rendered flush to the surface of the vessel. Set against a ground of irregular *leiwen*, the *taotie* on the vessel's body are composed of four confronted zoomorphs in profile which are separated by a smooth low ridge and marked at the forehead by a plain shieldlike crest. The animals, whose eyes are small plain round low-relief bosses set in canthi, are horned with large T-curls, while a hook of similar scale delineates the jaw. The bodies are single horizontal bands decorated with two spurred curls and a large swirling volute for a tail. Each has a single leg and foot, clawed, spurred, and tightly hooked. Decorating the upper face of the vessel and portions beneath the flaring mouth rim are four triangular blades. Each blade is composed of a single hooked inverted V-shape fitted at the apex of the triangle and a dragon with its head thrown back. With a long flat ribbon body broken only by a small hooked foot, the zoomorphs have large T-curl horns and curling proboscides as well as a curious curvilinear crest or excrescence along their snouts. Poised between the rising blades is a finely modelled animal head with protuberant eyes and large pointed ears, the canals of which are sunken and sharply curved.

The bronze was probably cast in a two-piece mold; it has later repairs at the legs. The patina is a fine, smooth, and often shiny surface with a dark olive green color. The corrosion is minutely granular in texture and mottled in appearance, ranging in color from blacks to light azurite blues, malachite greens, and cuprite reds. The interstices of the designs, particularly the *leiwen*, are filled with a light-beige earth.

The bronze is finely cast with two intaglio characters in a single column on the interior wall below the mouth rim. The inscription is visible over the rim of the *jiao* when viewed from the batlike animal head on the side opposite the handle. The graphs may be read *Tian Ding*, "Ancestor Ding."[1]

Known to exist as a vessel type from the Erligang phase, the *jiao* is, however, less common in both archaeological finds and collections than the ubiquitous *jue*, a fact that attests to its relatively limited manufacture and function during the Shang and Western Zhou. Although the two vessel types, *jiao* and *jue*, share similar features, such as the bladelike tripodal supports and the ovoid body, the *jiao*'s double points and lack of capped posts set it apart from the *jue*.[2] Excavated *jiao* are sometimes found with covers that are decorated with small simple loop handles and designs closely similar to those on the bodies.[3]

The zoomorphic creatures on the Saint Louis *jiao* are created in Loehr Style IV and are related in date and design to the Museum's *gu* with the bird inscription (No. 3; 28:1951). The shared design features of the two bronzes include the use of multiple parallel ribbons; single horizontal bands with spurred curls to comprise the bodies; T-curls and small plain shields at the flanges or ridges; tightly compressed hooked claws; and irregular *leiwen* as well as significant figures in relief. Subtle differences distinguish the designs of the *jiao* from those of the *gu*: small flourishes and elaborations of devices and motifs such as an increased number of parallel ribbons in the tail and an additional dewclaw on the foot or the encapsulation of the T-curl with an extra ribbon; all indicate a vibrantly developed and somewhat later state of this particular design.[4]

Selected Bibiography
"Additional Bronzes" 1951, 61, no. 4; *Handbook* 1953, 251; Bagley 1987, 211, cat. no. 24.

Notes
1. The character in the form of a standing man may have different readings: *da* (great) or *tian* (heaven). In this context, however, the graphs may read *Tian Ding* (as suggested in Kidder 1956, 50, pl. 10).

2. For a discussion of the development of the *jiao* vessel type see Bagley 1987, 208–14, especially 210, cat. no. 24.

3. *Jiao* may have covers in the form of birds, which were fabricated in later times (Bagley 1987, 210 n. 2).

4. Another wine vessel related to the Saint Louis *jiao* by its particular design is the older *jiao* in the Ashmolean Museum, Oxford (Deydier 1980b, 219, cat. no. 30) which is also distinguished by an acute V-shape at the mouth. The date of the Ashmolean bronze can be assigned to pre-Anyang on the basis of design (see Huber 1981b, 21). As a later manifestation of the design and its many elements as represented by the Ashmolean *jiao*, the Saint Louis *jiao* is transitional between the Erligang and Yinxu periods or, more likely, is early Anyang in date.

No. 5

Lei

13th–12th century BC
Shang dynasty
Height: 44 cm
Maximum diameter: 35 cm
Gift of J. Lionberger Davis
227:1950a

The tall shape, rounded shoulders, and footless base of the Museum's bronze are features of one of two types of *lei* found among bronzes at Anyang.[1] Therefore it is said that the Saint Louis *lei* comes from Anyang, and indeed the refined shape and designs of the piece do reflect the high metropolitan style of the late Shang. Although heavily restored, this *lei* merits discussion here for it represents a vessel type, shape, and course of design that significantly broadens the scope of the collection.

The *lei* is a wide-shouldered vessel with a plain cylindrical neck. Two loop handles with animal heads sit opposite one another high on the body, which curves inward at length to a flat base. There are four registers of decor beginning with two thick bowstrings around the bottom of the neck. Next is a band of interlocking doubled S-shaped curls bordered by a single row of circlets.

At the slope of the shoulder is a broader register of dragons with horns, hooked beaks, extended front legs and claws, and tails curled over their backs. These animals are rendered in multiple ribbons and intaglio lines of spurred curls. The eight dragons in the band are paired and addorsed; meeting in the center of the vessel, the confronting pairs are separated by a low-relief bovine head against a *leiwen* ground.

On the body, ten large pendant V-shaped blades stretch almost to the base. Each blade is composed of confronting dragons joined at their lower portions. The heads of the dragons are adorned with large spurred curls and parallel ribbons and end in axlike snouts. Their eyes are neat oval bosses with recessed pupils, and their feet are tightly balled with a single sharp spur. Along their bodies are spurred T-curls in intaglio. The motifs are set against a ground of fine straight parallel lines and *leiwen*.

The bronze has a dark brown patina which shows lighter muted gold colors and a mottled, somewhat pitted surface of black, malachite, and cuprite. The interstices of the design are filled with a gray-green corrosion and traces of beige earth.

A variety of *lei* vessel shapes can be traced to excavated Erlitou-phase ceramic wares and are represented in bronze form in numbers at Erligang-phase and later sites. The Saint Louis bronze exemplifies a sub-type that has a constricted mouth, a short columnar neck, and a relatively wide rounded shoulder. This shape of *lei* is comparable in proportions and size to notable examples from Anyang.[2] Both the Xiaotun and the Museum's *lei* have multiple pendant V-shaped blades and the noteworthy small band of interlocking, doubled, S-shaped curls. Rendered in Loehr Style IV, the *lei*'s major motifs, especially the blades, are, moreover, related to the elongated V-shaped designs found in ivory, horn, and bone carvings at Anyang.[3]

Close examination of the bronze reveals extensive but skillful repairs, including the replacement of areas with thin copper plate, the appliqué of bosses and corrosion products, and repatination. The animal-headed handles are probably not original. Stylistically, Shang-dynasty *lei* handles on this particular vessel type are thicker rings with low-relief modelled animal heads with prominent snouts. In high relief, the handles on the Saint Louis bronze resemble the lug handles on later Zhou vessels of similar shape and function.[4]

Selected Bibliography
Hoopes 1951, fig. 8; *Handbook* 1953, 249; Kidder 1956, 62–63, pl. 15.

Notes
1. Ma Chengyuan has described the types of *lei*: "one has a large mouth, short neck, wide shoulder, broad body, and tall foot-ring; the other is characterized by a small mouth, short neck, wide shoulder with two loops, tall body, a lid, and another loop on the lower side of the body" (Ma 1986, 199–200). Pre-Anyang *lei* have a different shape related to and often confused with the concave-shouldered *zun* (for a discussion see Bagley 1987, 144– 46, especially 146 n. 2, cat. no. 1).

2. For the white pottery jar, which is called a *hu*, but is actually a *lei* in shape, see Lawton 1972, 173–74, cat. no. 85, pl. 85. The bronze *lei* (R2076 from M238) is recorded in Li 1948, fig. 17b, pl. 1.2. Other *lei* with constricted mouths and flat bases were found at Xiaotun M18 (*Kaogu xuebao* 1981[4]:497–501) and Dasikongcun M53 (*Kaogu* 1964[8]:383), Anyang. The relationship between the Freer ceramic and the Xiaotun bronze was first noted by Loehr 1968, 100, cat. no. 42.

3. Compare the well-known carvings on ivory, horn, and bone from tomb HPKM1001 at Houjiazhuang, Anyang (Gao Quxun 1962, 1:269–passim; 2: pls. 202–6).

4. Compare the handles on *lei/ling* and *fou* vessels from the 8th century BC of the Spring and Autumn period of the Eastern Zhou dynasty illustrated in So 1995, 203–9.

Lei No. 5 (227:1950a)

No. 6

Ding

12th century BC
Shang dynasty
Inscribed: *Yin ban*
Height: 23.9 cm
Mouth diameter: 19.9 cm
Gift of J. Lionberger Davis
31:1951

Notable for its crisp decoration and fine proportions, this vessel is of a relatively modest scale but of impressive presence.[1] Details of interest on this bronze include the pointed and graduated flanges and the ambiguous rendering of the eye motif beneath each of the *taotie* tails.

The simple and direct shapes of the bowl, legs, and handles of this *ding* provide contrast to the crisp and dynamic forms that cover the surface of the vessel. The dense and compact decoration is enhanced by the constrained compartments on the face, which is divided into sections by six flanges. Each flange is of an overall even width but projects from the surface of the vessel at two different heights, the two levels relating to the two separate ornamented registers on the body. Marked by points and indentations which align with the upper and lower bands of decoration, the flanges are deeply scored with alternating horizontal and hooked T-shaped grooves. Two inverted U–shaped handles rise flush from the mouth rim; they thicken at their tops and flare gently outward. The beveled mouth protrudes as a slight everted lip from the body.

The zoomorphs, both large and small, are set in low relief against a *leiwen* ground of predominantly large, tightly wound, and often squared spirals. The top decorative register is a narrow band of twelve small dragons. Heraldically paired in the six sections defined by the flanges, the small zoomorphs face in the same direction as the larger dragon in profile in the main register directly below. Each dragon has a hooked and horned snout and a head similarly mounted with a sharpened crest, behind which is a larger curved and pointed horn. The animals' lower jaws are large and opened, sporting a single barbed square tooth, and their bodies have curled tails and two short legs, all detailed in curvilinear devices rendered in intaglio.

Below the large horns of the *taotie* are upright tightly curled C-shapes that rise high over their circular eye bosses and fanged jaws. Each *taotie* bears a tall frontal crest in an elaborate array of hooks and curls between its horns. The snout is squared and decorated with spurred curls in intaglio; similar recessed curling lines cover the body. The large curling tail is hooked and rises above a single clawed leg. Below each tail is a small eye motif with a single sharp vertical quill with hooks; unlike the dragons and *taotie* in low relief, the ambiguous shape and form of the eye motif is flush to the surface of the vessel.

The vessel stands on three short but well-proportioned cylindrical legs; each leg bears three pendant blades and an upper C-curl border in intaglio. The hooks and curls decorating the interior of each blade are asymmetrically arranged with devices of one side set slightly higher than the other. The border of recumbent C-curls is separated from the hanging blades and the vessel's plain base by two recessed lines.

A fine smooth matte patina of pale gray-green exists beneath a thin but compacted layer of bronze corrosion, encrusted mineral deposits, and traces of earth, occasionally highlighted by rust-colored patches.

The bronze was cast from a three-piece mold aligned on the legs and offset from the handles. The two-character inscription is cast in intaglio on the interior wall just below the beveled mouth rim and centered over a single leg of the vessel. A graph of two separate elements depicts a hand holding an object (*yin*) over a boat (*ban*).[2] Read as a single character, *ban*[3] in a few, somewhat longer inscriptions suggests that the possible meaning of the characters is that of a title or name.

Selected Bibliography
Hoopes 1951, fig. 4; *Handbook* 1953, 250; Kidder 1956, 48–49, pl. 9; Barnard and Cheung 1978, 798, no. 1345.

Notes
1. In shape and decor the Saint Louis *ding* is most closely comparable to a *ding* in Stockholm which is illustrated in Karlgren 1936, pl. 6, fig. A38. Another similar vessel is the tripod in *An Exhibition* 1940, cat. no. 4, pl. 6. See also a *ding* reproduced in Ecke 1939, no. 4.

2. The characters *yin ban* separately read "to govern" and "boat," respectively. There are four related inscriptions illustrated in Barnard and Cheung 1978, 798–99; a list of fourteen bronzes, including the Saint Louis *ding*, cast with the characters *yin ban* in their inscriptions is given in Bagley 1987, 294, cat. no. 49.

3. Zhou, et al. 1977, 1077–79, no. 2511.

No. 7

You

12th century BC
Shang dynasty
Height to top of lid: 19 cm;
to top of handle: 20.9 cm
Gift of Mrs. Charles Kunkel
in memory of her brother,
Dr. J. William Beckmann
576:1957

The swelling body of this *you* is elliptical in cross section, expanding gradually below the neck and narrowing to a constricted base and a splayed molded foot. The cover is a shallow dome with concave sides set over a plain interior collar which rises in oval form from the vessel's neck. The lid is surmounted by a stemmed six-lobed knob and four flanges that divide the top into quadrants along the short and long axes. The flanges, which are deeply scored with alternating straight and T-shaped grooves, follow the convex shape of the cover and bear blunt points at the ends. The four horizontal decorative registers on the whole of the vessel are similarly marked by sets of four flanges conforming in shape to the various concave and convex surfaces. A bail handle straddles the covered vessel and is attached to two lugs placed opposite one another along the short axis of the neck.

The ornamentation of the body is dominated by two large *taotie* featuring large recumbent C–shaped horns, small pointed ears, oval bosses with recessed pupils, and fanged upper jaws in low relief against a fine *leiwen* ground. The nostrils and nasal crests are decorated in intaglio with volutes, curls, and C-shapes symmetrically aligned on either side of the medial flange. The heads of the masks are attached to clawed feet and thick curling tails marked by spirals, C-curls, and pairs of confronted spurs. Flanking the *taotie* are pairs of inverted dragons in profile with gaping mouths, boss eyes, sharp horns, and curling snouts. Decorating the neck of the vessel are pairs of addorsed dragons with horns, curled tails, and two hooked feet, their jaws bearing a single sharp tooth. Slightly smaller versions of these dragons also appear, oriented in the same directions, in the horizontal register on the cover; here, the animals are separated on the short axis by a small *taotie* and low flange which can be hidden by the handle. The *taotie* decorating the top of the lid are essentially small-scale varieties of the large masks on the body. On the foot, the figures in this lowest register are distinct from the other motifs. Rendered in fine and broad thread relief, these small addorsed dragons are represented flush to the surface of the foot; only the small eyes are done in low relief. The inverted U–shaped handle is decorated by a lozenge which divides at midpoint the six dragons arrayed along the curved upper surface. At the ends of the handle are two animal heads with capped bottle horns, bulging eyes, and sawteeth.

In addition to the dark patina of the bronze, some areas are a smooth mustard yellow that may represent the original bronze color. The malachite corrosion is finely granular; the metal surface is pitted where cuprite is present. The corrosion color of the handle appears different from that on the body and lid which indicates a different bronze composition.

The oval cross section and low, swelling shape of the Saint Louis bronze conform to one of the four types of *you* that were made during the Shang[1] through the Western Zhou; all the ornamental designs, restrained flanges, and animal-headed strap handles are in accord with a late Shang date. Still, this *you* is distinguished from all others by the orientation of the paired and addorsed dragons that decorate the registers on the lid, neck, and foot. Animals facing away from center toward the sides of the vessel are infrequent in Shang bronze *you* designs.[2]

Furthermore, the covers of *you* vessels with oval cross sections and strap handles over the short axis usually sport prominent thick hooks aligned along the long axis.[3] The pair of short low vertical flanges on the sides of the Saint Louis cover are more often found on the lids of round-bodied *you*, which are tall cylindrical cannisters with bail handles,[4] or stout oval-bodied *you*, which are very similar to the Museum's *you*

in shape but have strap handles spanning the long axis.[5] The lack of these prominent hooks on the Museum's *you* and the substitution of low flanges appears to be a unique exception to the standard type.

Notes

1. Robert Bagley identifies four kinds of *you* made during the Shang dynasty: (1) tall, round-bodied; (2) double-owl; (3) round-bodied; and (4) oval-bodied. The oval-bodied *you* are sub-typed as having handles across either the long or short axis (Bagley 1987, 353–403, cat. nos. 61–70). Concerning the development of the stout oval-bodied *you* like the Saint Louis bronze, Bagley traces the shape of the vessel, the appearance of thick hooks on the cover, and the shift of handles from the long to short axis (374–77, cat. no. 64). He writes that the development of the specific features of the *you* "...do not exhaust the list of odd shapes—typological experiments—that might have been familiar to the casters who invented the oval-bodied *you*" (376). The variety of unusual vessel shapes that could have influenced the development of the oval-bodied *you* are indeed numerous and might have included covered oval-bodied *zhi* and vessels similar to the miniature covered vessel (49.135.lab) in the Metropolitan Museum of Art collection (380, cat. no. 65). Another uncommon Shang covered vessel for comparison is the early Erligang bronze in a private collection, St. Louis (illustrated in Lally 1994, cat. no. 44). This small bronze has a strange flat lid. The vessel is oval in cross section with a Loehr Style II *taotie* on both sides. It relates in conformation and design to the Metropolitan bronze by its corresponding recessed base, the pair of cruciform apertures and bowstring designs at the foot, and the two small projecting "tabs" at the neck which are aligned along the short axis.

2. Compare the dragons on the neck of the late Shang *you* in the Arthur M. Sackler Museum, Harvard University; the Sackler piece is also comparable in size to the Saint Louis *you*, standing 20.3 cm high and 15.2 cm wide (*Haiwai* 1985, 53, pl. 51).

3. The heavy hooks are derived from the beaks of double-owl *you* (Bagley 1987, 357, cat. no. 61) and usually occur on the covers of most stout oval-bodied *you* along their long axes. An unusual exception is the *you* in the Seattle Art Museum with hooks at the neck on the body of the vessel (*Haiwai* 1985, 22, pl. 22). Examples without hooks and comparable to the Saint Louis *you* include a *you* in the Shanghai Museum (*Shanghai Bowuguan* 1985, 179, cat. no. 32, pl. 32); and a *you* in the Hunan Provincial Museum (*Hunan Sheng Bowuguan* 1983, 169, cat. no. 42, pl. 42).

4. For example, see the round-bodied *you* in the Sackler Collection (Bagley 1987, 382, cat. no. 66); a similarly shaped bronze in the Asian Art Museum of San Francisco (Brundage Collection) (d'Argencé 1977, 48–49, cat. no. 16); and the early Western Zhou *you* in Kansas City (*Haiwai* 1985, 91, pl. 91).

5. Two examples of the *you* with handles over the long axis are reproduced in *Kaogu* 1963(12):646–47, figs. 1–2; and Deydier 1995, 107.

You No. 7 (576:1957)

No. 8

Gu

12th century BC
Shang dynasty
Height: 32.7 cm
Mouth diameter: 15.9 cm
Gift of J. Lionberger Davis
217:1950

The *gu* reached its full expression in the 12th century BC. The beaker becomes more attenuated in proportion, particularly at the waist, which lengthens and is more tubular. The tall aspiring form is enhanced by the more slender blades that rise to the mouth rim as well as a more tactile surface that is achieved by low-relief figures in all three primary design registers.

The flaring mouth and narrowed neck bear designs on a ground that is raised above the vessel's undecorated surface; this low but distinctive relief continues beyond the narrow plain band in higher relief at the waist. A second and broader recessed band decorated with two bowstring lines introduces the foot, the flare of which is undercut at a bevel. A high plain molding elevates the foot.

Four blades rise from a narrow band of four legless larval creatures that are silkworms. The dispersed and elongated elements of the *taotie* are in further relief above a crisp and highly articulated ground of *leiwen* patterns. Indeed, the same intricate and finely cast *leiwen*, described variously as Loehr V[a] and *leiwen*-relief style, fill the horns, snout, and other aspects of the *taotie* and all of the designs of the lower registers. The waist is decorated with paired *taotie*, each half of the mask separated from the other by raised flanges. At the foot, zoomorphic figures parade in a narrow band above larger *taotie* with horns of silkworms.[1]

The piece was cast from a two-part mold and bears no inscription. The patina is a smooth mottled gray-green color mixed with beige and pale blue areas, especially under the flaring mouth. Light blue azurite shows as granular corrosion deposits and occasional spots of cuprite can also be seen.

This *gu* is very similar to the other Saint Louis beaker (No. 9; 463:1956) in the sharpness of detail, shape, and form. Both vessels employ the low *leiwen* relief, the straight flanges at the waist, and curved pointed flanges at the foot. The three primary registers are also constant, as are the secondary registers at the bottom of the rising blades and at the top of the large *taotie* at the foot.

Despite the variations between the bronzes, these two *gu* most likely date from the same period, for they are in stylistic accord with late Shang *gu* from the archaeological site at Anyang, Dasikongcun.[2] The similarities and differences between these two Saint Louis *gu* may represent the late-Shang practice of burying sets of vessels in which pairs of *gu* and *jue* with similar or related designs were buried in tombs with the deceased.[3]

Selected Bibliography
Hoopes 1951, figs. 1–2; "Chinese Art" 1952, 71, fig. 23; Kidder 1956, 33–35, pl. 2; Watson 1965, 115, no. 85.

Notes
1. For a detailed discussion of the *leiwen*-relief style or Loehr Style V[a], see Bagley 1987, 247–50, cat. no. 36.

2. Compare the *gu* (53TSKM267:2) excavated from Dasikongcun, Anyang in 1953 (*Henan chutu* 1981, 50, cat. no. 306; 240, pl. 306).

3. The phenomenon of paired *gu* is exemplified by two undisturbed and contemporaneous late Shang graves at Jingjiecun in Lingshi, Shanxi excavated in 1985. Two well-provisioned tombs were dug out, each containing four *gu*, two vessels of one design and two of a related but different design. The archaeologists on site divided the *gu* into two groups on the basis of design: style I and II (*Wenwu* 1986[11]:1–18, pl. 2, figs. 9, 26). Of the Saint Louis *gu*, No. 8 corresponds to style I, and No. 9 to style II. The *leiwen*-relief styles on the excavated *gu* pairs have been compared with bronzes from Anyang tombs at Gaolouzhuang (57M8) and Dasikongcun (58M51) that date from the late Yinxu period III (*Wenwu* 1986[11]:17).

Gu

12th century BC
Shang dynasty
Inscribed: *X*
Height: 30.8 cm
Mouth diameter: 17.2 cm
Gift of Mrs. Charles Kunkel
in memory of her brother,
Dr. J. William Beckmann
463:1956

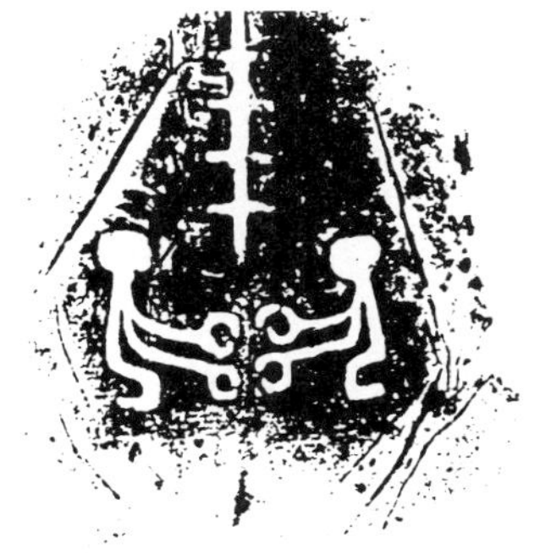

This *gu* is very similar to the previous Saint Louis beaker (No. 8; 217:1950) in its crispness of detail, shape, and form. Both vessels employ the low *leiwen* relief, the straight flanges at the waist, and curved pointed flanges at the foot. The three primary registers are also constant, as are the secondary registers at the bottom of the rising blades and at the top of the large *taotie* at the foot. The trumpet mouth and narrow neck display designs on a flat ground that is barely raised above the vessel's plain surface; this low but obvious relief continues beyond the narrow plain band in higher relief at the waist. A second and broader plain recessed band decorated with two bowstring lines and a pair of recessed cruciforms introduces the elevated foot; its flare and plain short skirt are sharply undercut at a fine knifelike bevel. Four blades rise from a narrow band of four silkworms. The disconnected and attenuated features of the *taotie* are in higher relief above a sharp and highly distinct ground of *leiwen* patterns. A similar complex and minutely wrought *leiwen* fills the horns, snout, and other aspects of the *taotie* and all of the designs of the lower registers. The waist is decorated with paired masks, their halves separated by raised flanges. At the foot, a band of silkworms lies above larger *taotie* with larval horns.

The piece, which was cast from a two-part mold, is covered in a mottled gray-green patina that is generally smooth and matte. The fine interstices of the *leiwen* pattern are often partially filled with a soft, beige earth; there are patches of blue-green corrosion and cuprite as well as dark carbonaceous blotches over the surface.

The foot of the vessel bears a cast inscription of a single undeciphered graph representing two kneeling figures facing one another. Between them is a spiked, pole-like device or standard.[1]

Although the shape and design of this *gu* are comparable to No. 8, there are subtle differences which affect its visual impact and presentation. Its foot is higher at the plain base and the width of the base is greater than half the diameter of the mouth. The high raised base has a broad smooth bevel that is slightly undercut, adding another edge to an already sharp profile. The proportions of the vessel, its registers, and designs are all compacted and uniform and rather more formulaic in their regularity. An indication of this somewhat reductive character is the absence of the *taotie* motifs in the rising blades at the neck and mouth. The secondary registers are also wider, incoporating additional lines of *leiwen* in their grounds. The blades rise from their bases in a constant width until their points. The flaring mouth is wider, more pronounced, and works in concert with the plain beveled skirt and broad elevated foot to accentuate the narrowness of the neck and waist. And while the rising blades of this *gu* are wide and bluntly shaped, there are no arresting *taotie* eyes to impede or interrupt the ascension of its finely wrought, aspiring forms.

In Shanxi province, paired sets of *gu* with similar designs have been scientifically excavated from tombs which date from the late Yinxu period and may have belonged to the Shang-dynasty rulers of the local region.[2]

Notes

1. A similar graph is illustrated in Zhou, et al. 1977, 250, no. 2102.

2. The site at Jingjiecun in Lingshi, Shanxi, which was excavated in 1985, yielded two well-provisioned tombs each containing four *gu*, two vessels of one design (style I) and two of a related but different design (style II); the archaeologists on site determined the two styles on the basis of design (*Wenwu* 1986[11]:1–18, pl. 2, figs. 9, 26). Of the Saint Louis *gu*, No. 8 corresponds to style I, and No. 9 to style II.

No. 10

Ding

12th century BC
Shang dynasty
Inscribed: *X Gui*
Height: 19.6 cm
Mouth diameter: 20.4 cm
Museum Purchase
22:1949

Poised beneath the broad lobes of the body, three plain cylindrical legs support this small *ding*. The vessel hosts two thin upright handles at the inward-slanting mouth rim. Below the everted lip of the mouth is a narrow band of twelve cicadas with bulbous eyes and heart-shaped bodies; the creatures are lined up head-to-tail in two groups of six which face in opposite directions. Three *taotie*, centered on the lobes and legs comprise the main ornamental device of the vessel. Raised round bosses, some with recessed dots for pupils, serve as staring eyes and focal points for the curving shape of the lobes. The foreheads of the beasts extend upward to form shieldlike crests; below are bluntly rounded snouts. Emerging from over each eye is a high, hooked, curling horn made of three thin ribbons; this shape is echoed by a curling tail with intaglio designs of hooks, curls, and spurs. The short narrow bodies of the *taotie*, which have relatively tight clawed feet, are similarly decorated with spurred volutes. Their jaws are filled with an upper file of sharply pointed teeth. Masks and cicadas are set against an irregular *leiwen* ground of circular, squared, and triangular spirals.

The overall patina is an even mint-green color of smooth texture overlaid with granular forms of malachite, azurite, and cuprite. Dark patches and smudges occur throughout, and the rim and interior show eruptions of corrosion. Copper red fills much of the finely rendered intaglio designs, highlighting the motifs against the light patination. The piece has been repaired.

A two-character inscription is cast just below the mouth rim in a single column which reads *X Gui*. Although the first graph is frequently seen in Shang bronze inscriptions, its meaning remains obscure.[1] The cyclical character *gui* is written in an earlier epigraphic form, dating to before the last two kings of the Shang dynasty.

The use of thin ribbons and varied intaglio line, tight clawed feet and upper jaws with barbed teeth, and irregular *leiwen* are all features of an earlier Loehr Style IV design developed circa 1300 BC.[2] Later forms of this early design are represented in the Museum's bronzes by *gu* No. 3 (281:1951) and *jiao* No. 4 (220:1950).[3] *Ding* No. 10 is an even later manifestation of this distinctive design. The design scheme of the trilobed *ding* is different from that found on round *ding*. *Taotie* on trilobed vessels are rendered across the full surface of each lobe and centered over each supporting leg, whereas on round tripods the *taotie* are set between the legs.[4]

Inscriptions are also formulaic. On lobed and round *ding* they are consistently cast on the interior wall over a single leg. They are properly read centered between the two upright handles above and two forward legs below. When inscribed *ding* are being read, the *taotie* on round *ding* are presented as full masks; on lobed *ding*, however, the animals face away from each other and cannot be read as full masks but rather as profiles of separate beasts.[5]

Selected Bibliography
Kidder 1956, 44–45, pl. 7; Barnard and Cheung 1978, 837, no. 1481.

Notes
1. The graph has been identified as a clan name (Chen Mengjia 1946), or as the verb *ju*, "to raise up" or "to offer" (Zhou, et al. 1977, 711–24, no. 2314). See Bagley 1987, 241–43, cat. no. 33.

2. Huber 1981b, especially 19–27.

3. Compare *jia* No. 2 (224:1950) for a similarly fanged upper jaw.

4. See Bagley 1987, 481, cat. no. 90 for a discussion of the design schemes on trilobed *ding*.

5. A *ding* has to be oriented in a certain way for its inscription to be read. That, in turn, affects the presentation of the surface design. The question is whether casting inscriptions in specific locations was generally independent of design schema, specifically the display of the full *taotie* on lobed *ding*.

No. 11

Ding

12th–11th century BC
Shang dynasty
Inscribed: *Shi Fu Xin*
Height: 21.8 cm
Gift of Mrs. Charles Kunkel
in memory of her brother,
Dr. J. William Beckmann
288:1955

The lobes of this *ding* are abbreviated and so produce a relatively shallow body. The beveled mouth rim is everted and flush with the outer sides of two inverted U–shaped handles which are rectangular in cross section. The body forms from beneath the lip into a trilobed vessel expressed in the slight swelling of each part. The main decor is comprised of three *taotie* in low relief, each centered on a lobe and marked by high protuberant bosses with recessed dots for pupilled eyes. The eyes are set in canthi over which are low-relief hooked brows detailed by a single intaglio line. The upper jaws of the beast are also rendered in low relief. The crest and jowls are marked by intaglio devices resembling the ancient Chinese graph for the left and right hands; the snout and nostrils are ornamented with curls and chevrons. Arching broadly above the mask are recumbent C-curl horns decorated with multiple hooks and outlines. The masks are flanked by upended dragons in low relief with small rounded bosses that have pupils, and curling tails. The motifs are set against an irregular *leiwen* ground.

The bronze has a granular patina of mottled gray-green which is occasionally splotched with cuprite and gray-white accretions.

The vessel bears a three-character inscription on the interior wall in two columns and reads *Shi Fu Xin*. The graph *shi* carries the meaning "historian" or "scribe" and is a possible clan name; the bronze was dedicated to a Father Xin. The inscription, which may be spurious, is cut into the bronze and must postdate the casting of the vessel. The work has been repaired.

The lobes and undercut jaws of the masks suggest heavy pendulous shapes and recall the udderlike ceramic models of this vessel type. Even the eye bosses bulge at a downward slant to enhance the weighty quality of the bronze. The decor is rendered in Loehr Style V and is marked by the prominent relief brows and nearly bosslike devices at the sculpted jaws which may or may not bear fangs. This decoration is a well known type used by both late Shang and early Western Zhou bronze casters.[1]

Selected Bibliography
Kidder 1956, 46–47, pl. 8; Barnard and Cheung 1978, 796, no. 1337.

Notes
1. For a detailed list of similar works, see Bagley 1987, 486–94, cat. nos. 93–95.

No. 12

Jue

12th–11th century BC
Shang dynasty
Inscribed: *Ya Xin*
Height: 21 cm
Width: 15.9 cm
Gift of Miss Leona J. Beckmann
in memory of her brother,
Dr. J. William Beckmann
344:1955

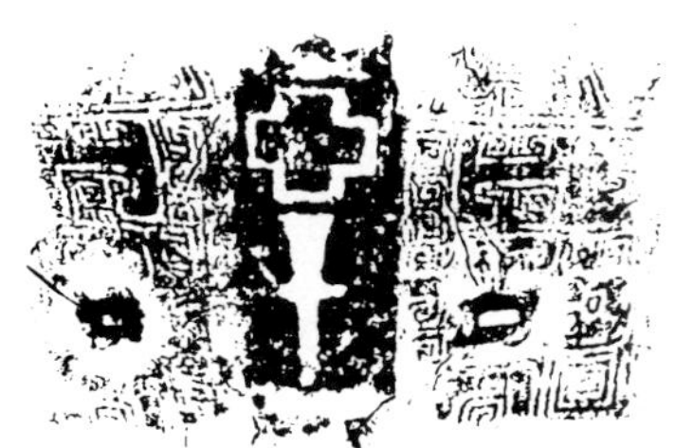

The spout of this *jue* is a long, U-shaped channel that arches higher than the point of the scooped and pointed tail on the opposite side of the vessel. Twin square posts, which rise flush from the exterior wall of the *jue*, are capped by nippled bosses. Elliptical in cross section, the body has a rounded base and is divided into four sections by the bovine-headed handle and evenly scored flanges beneath the spout, below the flaring point of the tail, and opposite the handle. Three slender legs of triangular cross section support the vessel; each leg tapers to a point and bears a shallow medial groove along one interior face. The top of each nippled boss is decorated with a line around its circumference and circlets in intaglio. The body decor is composed of two *taotie* with low oval bosses showing recessed pupils and brows of hocked C-curls. Spurred C-curls mark the horns and jaws with single hooked vertical quills framing each mask. Two pairs of blades with inverted V-shapes rise above the *taotie* to the finished mouth rim. A fifth and sixth rising blade decorate the undersides of the spout and tail. The features of the mask and the V-shaped motifs are created in plain ribbons against a *leiwen* ground.

The vessel was cast in a two-piece mold; mold joints are visible at the bottom of the two posts, along the spout, and on the handle, but all joints are well finished. Pseudomorphs of textiles appear in the corrosion underneath the base between the legs. The patina is an olive green overlaid with a thin mottled gray-green. A thicker layer of granular malachite corrosion coats some areas, especially under the spout, base, and legs. Occasional splotches of cuprite occur throughout the vessel.

The bronze is inscribed with two characters which read *Ya Xin*, in a single column on the exterior of the vessel just under the loop handle. The ya-shape or *yaxing*, a square notched at the corners, is a common graph of uncertain meaning, occuring in Shang and Western Zhou bronze inscriptions. In more complex examples, the *yaxing* may enclose a personal or clan name. Here in its simplest form, the *yaxing*, which is probably a clan name or noble rank, modifies the cyclical character *xin*.[1] A form of the *yaxing* occurs in the inscription of *jia* No. 20 (221:1950).

The *taotie* decoration of this *jue* is derived from the *leiwen*-relief style (also known as Loehr Style V[a]) in which the main features of the mask are in relief and decorated with fine *leiwen* patterns. The *leiwen*-relief style decorates Anyang bronzes that have been dated to the late Shang.[2] The later, reductive, and unadorned variation of the *leiwen*-relief style of the Saint Louis *jue* does away with the relief and the minute *leiwen* patterns on the *taotie* to produce a related but altogether different presentation of the basic design.[3]

Selected Bibliography
Kidder 1956, 52–53, pl. 11.

Notes
1. For a discussion of the *yaxing* see Cao 1980.

2. The *leiwen*-relief style or Style V[a] can be seen on *gu* Nos. 8 (217:1950) and 9 (463:1956).

3. For a close comparison with *jue* No. 12, see an example of the reduced *leiwen*-relief style of design on a *jue* (53YPM4:6) excavated in 1953 at Dasikongcun (*Henan chutu* 1981, 51, cat. no. 310; 243, pl. 310). No. 12 is also comparable to a late Shang *jue* in the Sackler collection without rising blades or raised flanges, but bearing a single low ridge with fine closely rendered lines (Bagley 1987, 196–97, cat. no. 19).

No. 13

Ding

12th–11th century BC
Shang dynasty
Height: 21 cm
Width: 14.8 cm
Bequest of Miss Leona J. Beckmann
28:1985

The smooth even surface of the Loehr Style IV decor and dark color gives this *ding* an extraordinarily compact character. The unremitting severity of its shape and form are softened by the plain large bosses in very low relief and by the rounded medial ridges that mark each lobe of the body.

Two inverted U–shaped handles which thicken at their tops rise from opposite sides of the slanted, beveled mouth rim. The top register is relatively large for the design type associated with this vessel shape[1] and is composed of small zoomorphs in smooth ribbons displaying horns, pupilled eyes, large jaws and snouts, and spur feet against a *leiwen* ground. In the main register below, each of three large masks with great circular eyes in low relief covers a lobe of the body. Their nasal crests are ornamented with hooks and spurs. The overarching horns are tight recumbent C-curls, nearly closed, and quite near the raised tails curling over the backs of the animals. The broad smooth features of the *taotie* are decorated with spurred and curling intaglio lines throughout. The background of *leiwen* in various design and sizes is finely done. The vessel is supported by three plain cylindrical legs.

The bronze, which is not inscribed, has a richly mottled dark brown patina. Cuprite and malachite are most evident, but there are also areas of earthy beige and other light-colored accretions. Present in small bits and patches, the overlying corrosion of reds and greens is visually cumulative and so darkens the overall color of the vessel. Moreover, the intaglio lines in areas are filled with a black granular substance.

Notes
1. For a comparable vessel and design see Bagley 1987, 480–81, cat. no. 90.

No. 14

Zun

11th century BC
late Shang dynasty
Inscribed: *Zi Zu Xin Bu*
Height: 26.9 cm
Mouth diameter: 21.1 cm
Gift of J. Lionberger Davis Art Trust
125:1951

The wide mouth of this *zun* is flaring and beveled to a narrow lip. The vessel is thickly columnar at its neck and bulges slightly at the short body, only to constrict somewhat and curve downward to the flaring molded foot. The foot itself is elevated from the ground by a plain footring and separated from the upper foot's flaring portion by a sharp undercut rim. High flat flanges divide the two registers of low-relief decor; the flanges are plain and have a nearly triangular cross section. Each flange on the body is convex to reflect the swelling shape of this section of the vessel; at the foot, they are slightly lower in height, pointed, and concave to follow and accentuate the contour of the flaring foot.

The *taotie* in the upper register on the body have recumbent C–shaped horns sculpted in two levels and ending in a sharp hook. The oval bosses, which are set in canthi adjacent to dimpled pointed ears, have recessed pupils in the form of a neat horizontal intaglio line. Below, the heavy hinged jaws bear four sharp fangs. On either side of the flanges are shieldlike crests decorated with fine intaglio curls and spurs and diamond-shapes. At the sides of the masks are large amorphous shapes in relief that suggest the bodies and tails of the animals. In the lower register, the *taotie* are smaller in scale and have flared nasal crests that are directly connected to the eyes, bodies, and multiple tails. The ears are contorted hooks, and the feet are composed of hooks and spurs. Simple bowstring bands are found in the pair at the neck and on the plain area just below the body.

The bronze was cast in a four-piece mold; the mold marks are visible on each of the flanges. The overall patina is a smooth pale gray-green with small areas of light olive and a darker network of medium- and dark-colored malachite. Overlying the patination is an accretion of chalky beige-white; at the foot are small spots of cuprite.

The vessel is dedicated to Ancestor Xin by four graphs cast in intaglio in a single column on the bottom of the interior. The inscription may be read *Zi Zu Xin Bu*. The first of the characters, *Zi*, is the graph for "son," which in this context may signify "the clan name of the Shang royal house."[1] The second and third characters read "Ancestor Xin." The last character, *bu*, is a clan name in the form of two stepping foot prints.[2] Usually, this graph appears in association with other characters; the individual *bu* graph of two stepping foot prints is rare and occurs, for example, on a late Shang *fangding* excavated in 1968 at Wenxian, Henan province.[3]

In terms of its shape and design, the Saint Louis bronze belongs to a small group of late Shang *zun* which have similar features.[4] All bear two lower registers of *taotie* and/or animal designs in smooth low relief against a plain undecorated ground; the upper portion, the neck and flaring mouth, are devoid of decoration and flanges except for the double bowstrings that are introduced close to the first ornamental register of the body. The shapes of these *zun* are broad and generous, but despite their capacity for holding ample amounts of wine, these vessels still possess an aspiring character which is enhanced by the elevated foot, the concave flanges, and columnar neck.

Selected Bibliography
Hoopes 1951, cover; *Handbook* 1953, 251; Kidder 1956, 36–37, pl. 3; Barnard and Cheung 1978, 759, no. 1213.

Notes
1. Bagley 1987, 486–89, especially 487, cat. no. 93.

2. Fang, et al. 1993, 104–5 cites a *zun* with the inscription of two feet and the Shang royal clan graphs *Bu Zi Zu Wu*, which may be read "[dedicated to] Ancestor Wu of the Bu/Shang royal clan."

3. Only the *fangding* is cast with *bu*; the two feet are placed left in front of the right as in the Saint Louis *zun*. The other three inscribed bronzes found in the Wenxian tomb have a different graph which looks like two feet placed side by side and aligned with two carpenter squares shaped like the capital letter L; the graph is read as *xi* (see *Wenwu* 1975[2]:88–91, figs. 2–4 in which the report makes no distinction between *bu* and *xi*; and Fang, et al. 1993, 123, which distinguishes the two graphs). The graphs *bu* and *xi* may be related; there seems to be little distinction made by the Shang inscribers between stepping feet and the modification of feet side by side with carpenter squares. A late Shang *ding* in the Sackler collection combines the stepping feet with carpenter squares (Bagley 1987, 490–91, cat. no. 94). The graph *xi* occurs in oracle-bone inscriptions where it is considered to be a clan name (for a discussion of the graph see Bagley 1987). In oracular inscriptions of the Shang, the clan name *xi* is usually preceded by the character for "dog," *chuan*. Thus, the *xi* clan may have been associated with the court office of the royal kennel (*Wenwu* 1975[2]:89). It is interesting to note that the kennel master undoubtedly had an intimate and precise knowledge of the ruler's territories. In addition to maintaining the king's kennels, the official would have been responsible in part for royal hunts and perhaps official tours of inspection which required an exact knowledge of the state's borders. Moreover, to some degree, the activities of this office may have been crucial in marking, confirming, or even changing the boundaries of the empire.

4. The dates for individual objects in this sub-type of *zun* vary from late Shang to early Western Zhou and thus may be considered a transitional style. A short list of these distinctive *zun* includes:

    a. an inscribed *zun* in the Sackler collections reproduced in Bagley 1987, 310–11, cat. no. 50; dated late Shang, the vessel is inscribed with two stepping feet with carpenter square borders

    b. a *zun* dated late Shang–early Western Zhou in the National Palace Museum, Taiwan (recorded in *Gugong* 1958, 1:74–75 and illustrated in 2:99; see also Chen Fangmei 1989, 34–35, pl. 37); it is interesting to note that the fifteen-character inscription on this bronze ends with a single three-toed left foot

    c. a *zun* (M50:4) excavated in 1973 from the Western Zhou tombs of Liulihe, Beijing (reported and reproduced in *Kaogu* 1974[5]:309–21, fig. 9.1), dated by vessel typology and epigraphic comparison to the late Shang and/or early Western Zhou (Yan 1975)

    d. an inscribed *zun* illustrated in *Henan chutu* 1981, 57, cat. no. 352; 279, pl. 352; this work is highly unusual in that it has a single register of decor at the body only; the foot has no designs or flanges but in all other respects conforms to the stylistic criteria that define this small group of *zun*.

No. 15

Fangding

11th century BC
Shang dynasty
Inscribed: *X*
Height: 23.3 cm
Width: 19.7 cm
Gift of Miss Leona Beckmann
in memory of her brother,
Dr. J. William Beckmann
108:1954

The four cylindrical legs of this *fangding* are slighty tapered; their upper parts bear horned *taotie*, each divided by short medial flanges. Rising flush from the beveled everted lip are a pair of inverted U–shaped handles that gradually thicken in width at their tops. The rectangular vessel, which is a little wider at the top than at the bottom, has a single scored flange down the length of each corner. All exterior sides of the body are decorated with a central horizontal panel of hooked and interlocking T-shapes[1] bordered at top and bottom by slanted and inverted V-shaped motifs. Above each central panel is a register containing two confronting dragons separated by a blunt, wedge-shaped stub in low relief. The animals are sinuous shapes with long upturned snouts, lower jaws, forward-curling tails, and spur feet on a *leiwen* ground.

On the longer sides of the vessel, the flanking and bottom ornamental panels show single upright or prone serpents with triangle-patterned backs and spade-pointed heads in slightly higher relief than the other low relievo designs. Surrounding the snakes on the *leiwen* ground are five bosses, most of which have three or four raised attenuated dots and others with round depressions only.[2] The flanking and bottom panels on the shorter sides of the body are decorated with diamond-back serpents.

The inscription is a single graph cast in intaglio on the interior wall of the vessel. In terms of design, the graph is essentially a square struck in the center with an X-shape whose small triangles fill the interstices. A pair of low, flattened arcs brace the sides of the figure and two lines of even length extend from the square at top and bottom. Evenly spaced between these linear extensions are two shorter lines.

The inscribed figure on this vessel is one of two variants[3] of a well-known graph which is cast on several other late Shang bronzes of differing vessel types and designs, including the small covered *zhi* in the Saint Louis collection (No. 21; 30:1985).[4] These other examples of the graph, which occur in both intaglio and relievo, differ slightly from the present inscription: all have a single short line evenly spaced between the parallel long lines extending upward and downward from the central square.

The reading and meaning of this graph in its present and common forms remains uncertain.[5] However, there are several other related pictographs which in their various associations with weaponry, ceremony, and currency or precious objects suggest that the graphs depict sites for the storage of weapons, for ritual, or for treasure. Thus the present graph may be a place name and may even represent a clan name.[6]

The Saint Louis *fangding* is one of a matched pair of bronzes: its mate is in the collection of the Palace Museum, Beijing.[7] The design of both vessels is unique.

Selected Bibliography
Kidder 1956, 58–59, pl. 14; Barnard and Cheung 1978, 848, no. 1516; Rawson 1990, 237, cat. no. 6.

Notes
1. The interlocking T-shapes in the central and side panels of the Saint Louis *fangding* are inverted and the mirror-reverse of a similar pattern on the *pou* vessel in the Sackler collection, described in Bagley 1987, 320–21, cat. no. 53. Given its many different configurations and permutations, this pattern varied to a certain degree from one bronze to another, and indeed can be found on many vessel shapes.

2. Shang and early Western Zhou bronzes with the snake and boss decor show clearly that each boss was meant to have three to seven raised dots, not depressions as in the case of the Saint Louis *fangding*. For a discussion of this process see Chase 1981, 113.

3. Another variant form of the graph, which has a single vertical line above and three vertical lines below, is reproduced in Zhou, et al. 1977, 802, no. 2356.

74

4. See Bagley 1987, 167–68, cat. no. 8, and Barnard and Cheung 1978, 849, nos. 1513–18 for other occurrences of the graph on bronzes. The graph on the Saint Louis *zhi* (No. 21) is cast in relief.

5. See Bagley 1987, 166–69, 169 nn. 2–3, cat. no. 8; and 472–75, 475 n. 2, cat. no. 88 for a detailed discussion of this graph and its relationship to other inscriptions. In its square form and in one peculiar form, in which the square of the pictograph is turned *en point* to form a lozenge or diamond-shape, the graph is read *zhu* with the meaning "to store things."

6. Louisa G. Huber notes that the graph is used in an oracle-bone inscription in the Kyoto Humanities Research Institute, "where it has been interpreted to signify the name of a tribe against whom an attack was being planned" (Huber 1981a, 78, fig. 5, detail). In some forms, halbards and arrows occur in or next to the graph (Barnard and Cheung 1978, 863, 865–67, nos. 1566, 1575–84); two kneeling figures present a vessel on or over the graph (Barnard and Cheung 1978, 776, 778, 782–87, nos. 1266–67, 1273, 1287–1303; 863, no. 1569); and cowry shells, the medium of exchange, and a vessel are depicted inside the graph (Barnard and Cheung 1978, 849, nos. 1519–21). The weapons, figures, and valuables are modifiers to the basic form of the graph. When associated with cowry shells, the graph in its Shang oracle-bone form has been determined to be a Shang clan name (Gao Ming 1992, 301–2).

7. For a reproduction of the other bronze, see Koyama, et al. 1975, no. 212.

Fangding No. 15 (108:1954)

No. 16

Fangding

11th century BC
Shang dynasty
Inscribed: *Xiaozi zuo Fu Ji*
Height: 21.3 cm
Width: 17.8 cm
Gift of J. Lionberger Davis
222:1950

The Saint Louis *fangding* is distinctive for its shape and style of decoration. Instead of the dense *leiwen* ground, multiple motifs, and elaborate structure of other Anyang-phase bronzes, this vessel adopts a simpler decoration which is a condensed and rather compacted version of the earlier bronze decorative mode of Loehr Style II. Found on bronzes dating from the 15th to 14th centuries BC, these are pre-Anyang designs in which the zoomorphs are delineated by flat ribbons for their bodies and plain bosses for their eyes. The effect of the decor is sculptural, a design in which the elongated and often flowing ribbons and curls of uneven width are complemented by generous, equally irregular, and visually complex negative spaces. In contrast, the design of spurred curls, S-shapes, and interlocking stemmed C-curls found on the Saint Louis piece is deliberately linear, depending on thin, even, regular lines to simply produce the wide forms that suggest the earlier Erligang-phase model but deny the richness of shadow and light of sculpted forms.

The pair of arch-shaped handles of this *fangding* incline slightly outward from the mouth rim and are thicker at their tops than at their lower ends. The handles meet the sloping rim flush to both the interior wall and the exterior face of the beveled everted lip. Four cylindrical legs, which are more thickly cast at their upper halves, support a body that is somewhat wider than it is deep, making a rectangular cross section. The central panel on each side of the body is plain. Above it is a decorative frieze of a *taotie* composed of two confronting animals in profile with rounded rectangular bosses for eyes. The profiles are separated from one another by the low blade of a vertical flange. The head and body of each zoomorph are made up of spurred S-shapes and interlocking stemmed C-curls and a single pointed quill. The central *taotie* is flanked by two square panels made of similar elements which create animals in profile facing away from the center mask. When viewed from a corner of the vessel, full masks are created by the squared profile of one zoomorph meeting that of another. Below the ornamental register and cupping the central blank panel is a broad angular U-shape filled with straight rows of raised bosses of uniform shape and size. The crowns of the legs are decorated with masks and triangles rendered in intaglio lines with spurs and interlocked C-curls. On each leg is a short incipient flange which is aligned with the corner of the vessel and serves to separate confronting animal profiles.[1] Below the mask is a set of three pendant triangular blades. The handles are simply decorated with two intaglio lines that follow the curve of the arches.

The entire vessel has a fine smooth patina of gray-green overlaid in areas with a granular corrosion of malachite and irregular black patches and orange-red earth.

The inscription of five characters in a single vertical column cast on the interior wall of the vessel reads *Xiaozi zuo Fu Ji*, "Xiaozi made [this vessel for] Father Ji."[2]

In concert with the archaistic style of ornamentation, the shape of the Museum's piece is also based on two early Erligang prototypes of the *fangding* vessel type which occurred in both large and small sizes.[3] While there are many differences between the excavated bronzes and the Saint Louis vessel, the basic shape, design, and motifs are strikingly similar, especially to the very large prototypes: the squarish body supported by thick short legs; the single high register with a full *taotie* and flanking animal profiles; the plain central panel; and the rows of bosses.

One difference between the Erligang-phase vessels and the Saint Louis *fangding* is the construction of the handles, which on the earlier bronzes are hollow horseshoe shapes strengthened by one or two inverted U–shaped ribs that are recessed, parallel, and concentrically arranged within the handle.[4] The two plain intaglio lines on the

thick solid handles of the Museum *fangding* are decorative echos of the comparative-
ly thin and hollow cast handles of the earlier pre-Anyang models.

In terms of archaeological finds and known collections, the Museum's *fangding* is
thus far unique and comparable to only three other late Shang vessels[5] which share a
similar shape, elements of design, and miniaturistic scale.

Selected Bibliography
Hoopes 1951, fig. 5; *Handbook* 1953, 250; Kidder 1956, 56–58, pl. 13.

Notes
1. The small masks on the upper portions of the legs are rendered in finer lines, producing broader and
rather pneumatic forms. Derived as they are from the schema in the highest ornamental register of this
vessel, the designs of these small masks on the leg share a similar quality of abbreviation and compression; however, despite the essential curvilinear features of the design, the masks appear embryonic in
nature and so differ from the animals in the upper registers on which they appear to be based.

2. The same inscription is cast on a *ding* in Chen Mengjia 1977, pl. 11. See also Loo 1941, cat. nos. 23,
30 for a similar inscription, *Xiaozi zuo Mu Ji*.

3. The prototype for the *fangding* is most notably represented by the two large bronzes unearthed at
Zhengzhou in 1974, which are also among the earliest known *fangding*. Both *fangding* were found in
situ, side by side, in an excavation in September 1974 at Zhangzhai Nanjie, Zhengzhou in Henan
province and are now in the collection of the Henan Provincial Museum. The larger of the two vessels
is 100 cm high by 62.5 cm wide and weighs 82.4 kilograms (181 pounds, 4 ounces); i.e., the
Zhengzhou works are squarer in shape and more massive than the Saint Louis bronze (*Wenwu*
1975[6]:68, figs. 10–11; *Henan chutu* 1981, 6–7, cat. nos. 34 and 35; 38–40, pls. 34–35). In 1982, two
more large *fangding* (H1:2 and H1:8), both measuring 81 cm high and buried side by side, were excavated at Zhengzhou from a single pit (*Wenwu* 1983[3]:49–59, figs. 3, 18–19, 25). Of considerably
greater comparative value are the two small *fangding* excavated in Beijing in 1977 that also date from
the Erligang phase. The vessel for which measurements are given is a mere 14.2 cm high, including the
handles, and rectangular in shape. While this early work is miniaturistic in scale, it is more similar in
size and shape to the Saint Louis bronze than the larger Zhengzhou examples (*Wenwu* 1977[11]:1–8,
pl. 4.1, figs. 21–22; and also Zou 1980, especially 264, pl. 40.1). A later example of the small *fangding*
is the single vessel (M5:834) excavated from the tomb of Fu Hao dating from the Anyang phase, Yinxu
period II. While the larger *fangding* examples of the Erligang phase provide the design scheme for the
Saint Louis piece, it is the smaller *fangding* of the Erligang and earlier Anyang that provide the scale.

4. See Lienert 1979, 2:240, no. 81 for an illustration of a hollow handle. For a detailed explanation of
the casting of the Zhengzhou *fangding*, including their handles, see Robert Bagley's discussion in
Bagley 1980, 108, cat. no. 11; and Bagley 1987, 62 n. 217, cat. no. 49.

5. The inscribed *Geng Shi fangding* (82M1:44) excavated in 1982 from Tomb 1 at Xiaotun, Anyang is
the closest comparison to the Saint Louis bronze. It stands 22.6 cm high and 18 cm wide (*Yinxu* 1985,
232, fig. 85, no. 1). Compare a *fangding* in the Tokyo National Museum collection; and a *fangding* in
London, formerly Boisgirard and Heeckeren collection, Paris (Deydier 1980b, 20, pl. 2; also Deydier
1980a, 12–13, cat. no. 1). All of the bronzes are of similar size, shape, and design and differ in only the
most minor details. There is, however, no identical bronze with the same design scheme as the Saint
Louis bronze, which has zoomorphs with plain eyes and small vertical flanges on its legs, while all
known comparative bronzes have eyes with recessed pupils and no flanges on their legs.

Lihe

11th century BC
late Shang–early Western Zhou
dynasty
Inscribed: *Yi Fu X*
Height: 31.6 cm
Gift of Mrs. Charles Kunkel
in memory of her brother,
Dr. J. William Beckmann
287:1955

As is clear from its prominent spout and ample proportions, this *lihe* is a vessel used to contain and pour liquids. Its form revives and so finely adapts aspects of earlier Shang bronze in body type, decoration, and style that it conforms readily in presentation and character to the distinctly archaistic trends in bronze design that occurred in the 11th century BC.[1]

The vessel is supported by three tapered legs that extend down from the full, swelling, trilobed shape of the body; the rearmost leg provides the centerline for the handle, connecting rings, and spout. The vessel has a high-angled tubular spout located on the shoulder opposite a closed curved handle. The constricted neck bears an everted and beveled mouth into which fits the low domed cover. Rings on the lid and body are connected by a double-ringed link. The lid is surmounted by a hollow annular knob with a slightly flaring profile and is decorated with an intaglio frieze of ovals and *leiwen* curls.

The plain ovals alternate with blocks of spurred curls; directly below this continuous design is a tight row of neat circlets. A similar but slightly wider band of ovals and curls decorates the shoulder of the vessel proper where the band is finished with an upper and lower border of circlets. This second frieze is further distinguished from the decor on the lid by two eyes with recessed pupils flanking the spout.

The spout itself is decorated with four rising blades and a small band of C-curls. The handle, which joins the body below the ornamental register, is decorated at its top with a modelled bovine head; the remaining length of the handle is covered with spurred curls and hooked devices enclosed by a border of two intaglio lines.

The vessel has a smooth dark brown patina with muted gold highlights on the lid and handle. Areas of the cover and body are lightly encrusted with small hemispherical eruptions of pale malachite that aggregate as heavier corrosion and spots of light-blue azurite. The surface is lightly speckled with cuprite and deposits of fine earth.

A three-character inscription is cast twice; once, in a single line on the body underneath the animal head of the handle and a second time inside the lid. The lines are composed of two commonly known ideographs: *yi* (the second cyclical stem or *tiangan*) and *fu* (father), and one undeciphered graph (*X*) which depicts a field between a pair of hands. If the characters are read in sequence as cast from top to bottom, they read *Yi Fu X*. This order, however, reverses the usual sequence which would read as a dedication to *X Fu Yi*, "Father Yi of the **X** Clan."[2]

Because of its trilobed shape, which is called a *li*, the Saint Louis bronze can be most accurately described as a *lihe*. The *li* first appears in neolithic pottery as a pendulous, udder-shaped container: three separate vessels joined at a common mouth and spout that could heat the liquid contents more rapidly because of increased surface area. In the Shang dynasty, the full, pneumatic shape of the *li* was adapted to other vessel types, including the *ding*, *jia*, and *he*. *Lihe* of the middle Shang have divided tripartite bodies with straight and erect spouts.[3] Later vessels like this one retain the high angled spout but their trilobed parts are conjoined so that one lobe flows smoothly and without separation into the other.

Given the *li* shape's neolithic and middle Shang prototypes in both ceramic and bronze, its revival during the late Shang and early Western Zhou is a conscious archaism. In the case of this vessel, the archaistic quality of the lobed vessel shape is complemented by the equally archaistic rendering of the decor, which in treatment resembles the curling T-shapes and flat ribbons of the middle Shang. While their model is Loehr Style II of about the 15th–14th centuries, here the ribbons are modi-

fied; they are smaller and condensed and appear contained in blocks which alternate with ovals. The reduction of scale and the compression of designs into geometric shapes anticipates the heightened repetition and patternization of motifs characteristic of the succeeding Zhou dynasty.

Selected Bibliography
Kidder 1956, 42–43, pl. 6; Barnard and Cheung 1978, 814, no. 1396.

Notes
1. The dates of similar *lihe* are variously given as late Shang or early Western Zhou on the basis of either inscription or archaeological excavation: see Bagley 1987, 304, fig. 49.22, cat. no. 49; and *Kaogu xuebao* 1977(2):99–130, pl. 8.2. Works comparable to the Saint Louis *lihe* include the vessels in the Sackler collection (illustrated in Rawson 1990, 662, cat. no. 112) and the British Museum (Rawson 1990, 665, fig. 112.5).

2. Kidder 1956, 42; and Barnard and Cheung 1978, 814, no. 1396.

3. See d'Argencé 1977, 20–21, cat. no. 2 for middle Shang *lihe* with divided tripartite bodies.

Lihe No. 17 (287:1955)

No. 18

Tiaose qi

11th century BC
late Shang–early Western Zhou
dynasty
Height: 17 cm
Gift of J. Lionberger Davis
33:1951

The *tiaose qi* is among the rarest of bronze vessels. Used in ancient times as a container for red, green, white, and black pigments, this vessel type occurs in pottery and stone as well as bronze forms.[1]

The Saint Louis *tiaose qi* is a tetrapod with narrow, bluntly pointed posts for feet. The posts flare up into the rounded bases of four separate cylindrical containers of equal size which are joined together into a squarish vessel. The rims of the four open cylinders rise a bit above the surface of the square top, which is pierced with a round aperture of a diameter similar to the cylinders. There are two small closed-ring handles surmounted by bovine heads on opposite sides of the vessel. This pair of sleek animal heads is similar in type to the more massive and fully modelled bovine head that juts from the S-curve neck at the front of the *tiaose qi*. The large head is composed of two sweeping figured horns, a pair of pointed dimpled ears, and flaring nostrils that punctuate a curiously blunted muzzle. The eyes are simple bosses set in canthi and slightly raised brows. The elongated neck is decorated with a single long row of large cowry shell motifs. A band of two spurred C-curls on thick stems ornament the upper portion of each cylinder. Below this band are three triangular blades with points and curls resembling an anthropomorphic pendant masklike figure. With the exception of the animal heads, all decor is rendered in a somewhat cursory, uneven, and shallow intaglio line.

The bronze has a patina of a rich leaden color overlaid with patchy areas of granular corrosion of pale green malachite and spots of red cuprite. The smooth malachite along the neck is dark green and contrasts with rough areas of rust throughout the vessel. Red pigment coats the interior of the second cylinder clockwise from the large animal head.

At the early Anyang sites of Houjiazhuang and Xiaotun, a few *tiaose qi* made of marble and pottery have been excavated. These stone and ceramic types are decorated with incised plain bowstring designs encircling the round low bowl-like shapes. The vessels have four or five round wells for pigments sunken in their flat tops.[2] A rare vessel type to begin with, only a very small number of bronze pigment containers have been excavated and, like the Saint Louis piece, they all take the shape of four cylinders forming a square, a shape which appears to be special to the bronze medium. A bronze animal-headed *tiaose qi* was excavated at Qishan Hejiacun Xihao, Shaanxi; it bears a bovine head with swept-back horns and similar heads on the two ring handles, very much like the Museum's piece.[3]

Selected Biography
Kidder 1956, 74–75, pl. 20; Chêng 1965, 246, pl. 4, fig. 6.

Notes
1. For a discussion of the *tiaose qi*, see Chêng 1957, 34 and Chêng 1965, 239–250. The *tiaose qi* was once considered a "ceremonial lamp" (see Kim 1940 and Allen 1950, 134, fig. 2).

2. For descriptions of these marble and pottery *tiaose qi* see Gao Quxun 1962, 1:99–100; 2: pl. 100, figs. 3A–B, pl. 102, fig. 3; and Li 1956, 62, xxxvii, cat. nos. 210D–E.

3. The *tiaose qi* from Qishan Hejiacun Xihao measures 15 cm in height, which is very close to the Saint Louis piece: see *Shaanxi chutu* 1979, no. 158. A plain, rectangular undecorated openwork bronze *tiaose* (72AGM82:4) was excavated in 1972 from tomb AGM82 in Yinxu, Anyang. The late Shang vessel was misidentified and mislabeled as a "*qizuo*" or "implement stand." (*Henan chutu* 1981, 33, cat. no. 200; 158, pl. 200). Comparable works include a superbly cast *tiaose qi* with extensive intaglio designs in the Grenville L. Winthrop collection of the Harvard University Art Museums; and a severely plain piece with an animal head in the Ashmolean Museum, Oxford, formerly in the Sir Herbert Ingram collection (Davis 1956, fig. 12).

No. 19

Fangyi

11th century BC
late Shang–early Western Zhou
dynasty
Inscribed: *X Fu Yi*
Height: 21.3 cm
Width: 13.8 cm
Gift of J. Lionberger Davis Art Trust
127:1951

Fangyi first appeared as a distinctive vessel type in the Shang dynasty and continued to be made for successive marquis of the State of Jin into the early Spring and Autumn period (770–ca.670 BC).[1] In general, this type of vessel is completely covered with decoration.

The lid of this unusual *fangyi* is surmounted by a knob on a square stem that meets the cover along the topmost ridge; the shape of the knob replicates in miniature form the pyrimidal shape of the lid. The cover, which is fitted into the interior of the rim, overhangs the short, angled lip rising from the top ornamental register on the vessel. The body of the container is roughly square in cross section and wider at the mouth rim than at the recessed base. The beginning of the hollow foot is marked by a shallow groove running above the lower decorative band.

The vessel and its cover are decorated in three registers, all of similar design. The top decorative band on the body consists of two confronting dragons. The animals are separated by an inverted triangle composed of large C-curls in intaglio and a smooth low-relief vertical ridge, the sides of which are scored with minute indentations. Each zoomorph has a vertically aligned angular boss with a recessed dot for an eye; its body has a single curling horn and tail and two feet. The large lower jaw is punctuated by a single spurred tooth. At the far edges of the band are low finlike flanges marked with fine regular lines. The animals are rendered in thin ribbons of fairly uniform width by even intaglio lines. The design of the shorter ornamental register on the foot is similar to that above it, differing only in its slightly compacted quality and triangular crest. The crest and low-relief ridge are interrupted by a small arch that accentuates the placement and shape of the inverted V–shaped notch in the foot rim. On the lid, the bosses are much lower in relief and have a horizontal orientation as opposed to the vertical alignment of the bosses on the body. The space between the two decorative registers of the body is plain except for a thin bowstring line in relief. The knob on the lid bears two intaglio *taotie* composed of recumbent C-curls for horns, ovals for eyes, and a distinctive spurred curling line for the snout. The other sides of the knob are decorated with triangular spurred curls.

The cover and body of the vessel have a smooth light-colored gray-green patina which is overlaid with a fine granular and pitted crust of pale malachite that also partially fills the intaglio designs. Beneath the malachite crust are occasional spots of red and orange cuprite and irregular areas of black color and reddish earth.

The vessel bears an inscription of three characters in two columns, on both the interior base of the body and the inside wall of the cover. The undeciphered graph in the left column may relate to the character and surname *Wu*.[2] Reading left to right, the inscription may be read *X Fu Yi*, "X Father Yi."

Only in rare instances is the bronze plain or sparsely ornamented like the Saint Louis *fangyi*; there are only a few comparable works which date from the early Anyang to late Anyang phases.[3] Early examples of the vessel have a foot rim which is opened on all four sides by a broad wide arch.[4] In late Shang *fangyi*, the opening is a very small arch on the broad sides of the vessel or a mere inverted V–shaped notch on the short sides.[5] The use of the notches, as opposed to the arches, on the Saint Louis *fangyi* indicate a late rather than early date for the piece.

The Museum's *fangyi* also displays a number of other late Shang features. The short, sharp-edged, finlike flanges scored with fine lines[6] and the style of thin ribbons derived from earlier bronze design, specifically Loehr Styles II and III, indicate a late Shang or transitional late Shang–early Western Zhou date for the vessel.[7]

Selected Bibliography
Kidder 1956, 40–41, pl. 5; Barnard and Cheung 1978, 731, no. 1128.

Notes
1. An early *fangyi* of Shang date was discovered in Anyang (Li and Wan 1972, pl. 13). Later examples of the vessel type have been found in recent years. In 1993, a late Western Zhou *fangyi* (I11M63:76) called a "funerary piece" was discovered in the tomb of the secondary consort of the Marquis Bang of Jin (*Wenwu* 1994[8]:4–21, fig. 24.1). In 1994, an elaborate covered *fangyi* (M93:51) of the early Spring and Autumn period was excavated from the tomb of the Marquis Wen of Jin (*Wenwu* 1995[7]:4–39, figs. 43.7, 45).

2. The undeciphered graph in the left column is similar to the well-known character and surname "*koutian*" *Wu* which is usually formed by the elements *kou* (mouth) and *tian* (sky or heaven) *Wu*. In numerous ancient examples of the character *Wu*, the *tian* element takes the form of a human figure with its head angled to the left and the *kou* element inserted at its neck. The graph on the Saint Louis *fangyi* possesses the structure of the *Wu* character, but the inverted triangular element at the neck of the figure does not resemble the usual *kou* or "mouth." Moreover, the triangle in the graph is elaborated with a short interior line parallel to its base. *Kou* or "mouth" does occur, albeit infrequently, as a triangle; indeed, a triangle is substituted for *kou* in a variant character of *Wu* cited in the *Shuo Wen* (Xu Shen 1975, 2:10:2a; see also Zhou, et al. 1977, 116, 152, nos. 2025, 2042). There is an increasing likelihood that this character should be read as the familiar "*koutian*" *Wu*. Note also the reference to a "*Wu fangyi*" in Fang Shuxin et al. 1993, 764–65.

3. The following four bronzes are distinctive for their restrained use of decoration and for the large surfaces devoid of designs:
   a. The earliest example is a plain *fangyi* with bowstring designs, wide arch openings at the foot, and two animal heads in high relief (Goepper 1972, cat. no. 1, pl. 1). This work probably dates to the 13th century and is comparable to the *fanghu* formerly in the Werner Jannings Collection, Palace Museum, Beijing, the top portion of which is plain with three bowstrings and relief animal heads (see Consten 1958, pl. 1). A plain three-bowstring design is found on a 13th-century square stand (76AXTM5:850) from the tomb of Fu Hao, consort to King Wu Ding (*Henan chutu* 1981, 32, cat. no. 194; 155, pl. 194).

   b. A later Shang example of a *fangyi* formerly in the H.G. Oeder collection (Karlgren 1937, p. 69, pl. 54) is discussed in relationship to the Saint Louis bronze in Karlgren 1962), pls. 1–80.

   c. The latest comparable work in date to the Museum's *fangyi* is a very similar bronze of late Shang–early Western Zhou with animal-headed lug handles and rings, illustrated in the Parke-Bernet auction catalogue *Important Early Chinese Dynastic Bronzes* 1948, cat. no. 1000.

   d. Compare the *fangyi* (M9:6) excavated in 1991 at Hougang dating from the late Yinxu period IV.

4. Max Loehr identified the arch as a means of comparing early and late *fangyi* (Loehr 1968, 92, cat. no. 38). For a detailed discussion of the arch on *fangyi* see Bagley 1987, 430, figs. 77.7–9, cat. no. 77.

5. For a bronze dating to the reigns of the Shang kings Wu Yi and Wen Ding (Yinxu period III) compare the late Shang *fangyi* with inverted V–shaped notches (M269:22) discovered in 1984 at Qijiazhuang, Anyang (*Kaogu xuebao* 1991[3]:325–52, pl. 12.2, figs. 10.1–2).

6. Compare a late Shang *gu* with similar low flanges (*Henan chutu* 1981, 40, cat. 238; 189, pl. 238).

7. Note that the small *taotie* on the knob of the Saint Louis piece is very similar to small animal mask designs associated with Western Zhou bronze seals (*Wenwu* 1995[12]:76–77, figs. 4.2–3).

Jia

11th century BC
Shang dynasty
Inscribed: *Guisi Wang xi Xiaochen Yi*
bei shipeng yongzuo Mu Gui zunyi
wei Wang liusi yongri zai siyue Ya Yi
Height: 45.9 cm
Mouth diameter: 23.5 cm
Gift of J. Lionberger Davis
221:1950

The stately presence of this *jia* and its lengthy inscription have long attracted scholars and collectors. The statesman and connoisseur Duanfang (1861–1911) was the first to note the vessel in a catalogue of his collection of important inscribed bronzes, the *Taozhai jijin lu*, published in 1908. At once vigorous and refined, the *jia* offered not only evidence of the fine art of late Shang bronze casters, but by its inscription the bronze also exemplified the relationship between king and subject as well as the model response to royal favor. Then as now, the aesthetic, stylistic, and documentary character of the Saint Louis vessel was irresistible.

The Museum's *jia* is a formidable and robust bronze vessel of relatively large scale. The body is sectioned into three large lobes with legs forming a tripod: each lobe slopes from the broad neck and then swells, the outline then curving down to a cylindrical foot. Wide pointed arches are formed below the vessel by the legs and the meeting of the lobes. Truncated flaring cones cap rounded twin posts that rise above the mouth rim from the interior wall. The posts are flat and flush to the finished everted lip. A large closed handle of circular cross section is cast to the neck and third lobe of the body.

The *jia* is decorated with two bands of *taotie*. The masks on the neck are comprised of elongated animals, each with multiple quills, legs, and a spiral tail curling over its back. Surmounting its oblong, pupilled eye is a horn created from a recumbent spurred C-curl with a medial stem. Hooked quills rise vertically from the back above numerous hooks for legs and feet below. Decorating the shoulder is a register of long multi-legged *taotie* composed of many parallel ribbons, whose oval eyes with recessed pupils are also found on the neck. There are no quills in the register; instead, there are paired hooks and T-curls all along the body. On the neck and shoulder both bands are divided by short ridges in low relief. Along the bottom of each lobe of the body is a V-shaped motif made of double bowstrings in low relief, the bowstrings rising from above the foot to outline the triple pointed arches beneath the vessel. The *taotie* stand against a ground of fine regularly formed *leiwen*. The handle is cast along the median of one lobe and joins the body at the neck and just midway between the shoulder and foot. The handle issues without decoration from the jaws of a bottle-horned animal head with pointed ears, bulging eyes, and fangs. The head is minutely detailed: the eyes are set in recessed canthi below arching hooked eyebrows; the tops of the caps of the bottle-horns are decorated with symmetrical patterns and neatly finished curls; these small horns mirror the larger conelike caps on the posts on the rim above. Each large cap is decorated with a band of spurred curls and six finely ornamented triangles which point to a pair of thin bowstrings.

The surface of the vessel is smooth with very fine pitting. The darker areas are a deep olive green encrusted with spots of malachite, azurite, and cuprite corrosion of nearly uniform size. In some large sections, under the handle and at the neck for example, the surface has a lacquerlike luster which contrasts with the matte finishes of the corroded areas.

The Saint Louis bronze inscription is relatively long. Early bronze inscriptions consisted of a few characters, usually clan signs or the names of ancestors to whom the vessels were dedicated. During the late Shang, some bronzes, such as the Saint Louis *jia*, were cast with much longer texts that recorded the award of royal gifts to court officials for their loyal service to the king. The ruler expressed his favor by presenting strings of cowry shells, the Shang form of currency. The official then used the

funds to commission the casting of a bronze vessel which not only commemorated the royal gift, but also honored an ancestor.

The Saint Louis *jia* has a long inscription of twenty-seven characters in two columns of equal length, cast on the side of the vessel beneath the handle.[1] The characters are clearly and precisely written, crisply cast in a deep intaglio on the curved surface of the body under the animal-head handle. The inscription may be read as:

> [On the day] *guisi*,[2] the king[3] bestowed upon the Xiaochen[4] Yi[5] ten strings of [cowry] shells[6] [which Yi then] used to make for Mother Gui[7] a sacred vessel. This was in the king's sixth *si*,[8] during the *yongri*[9] [ritual cycle], in the fourth month. *Ya*[10] *Yi*.[11]

The inscription on the Museum's *jia* conforms to a formula found in both oracle-bone and bronze inscriptions of the late Shang. These oracular and dedicatory conventions were identified in 1956 by Chen Mengjia who recorded fifty-four comparable inscriptions on ancient bronze vessels, the bones of animals, and oracle bones used in divination. The Saint Louis piece is included on this list of ten extant and recorded bronzes and the inscribed and inlaid bones and skulls of three kinds of trophy animals, tiger, rhinoceros, and deer, killed on royal hunts.[12] The distinguishing elements in all of the writings involve the manner of dating the inscription used by the scribes: a sequential, literary, and stylistic prescription which included the cyclical day, the king's *si*, the minor ritual cycles of *Xie*, *Yi*, or *Yong*, and the use of the characters for the preposition "on [a certain month]" and the phrase "this was in [the king's sixth *si*]."[13] Xu Jinxiung identified other characteristics of late Shang writing. In the last phases of the period, passages refer expressly to the rulers and sacrifices to kings and queens that were made on the same cyclical days as the cyclical names in their posthumous titles.[14] This court practice is reflected in the present inscription, where the king's gift is commemorated and the dedication of the bronze to Mother Gui is made on the *gui* [-*si*] cyclical day.

In his study of oracle-bone inscriptions, Xu Jinxiung also noted rather precise stylistic elements in late Shang composition and calligraphy. These same features of style and configuration can be applied to the inscriptions of bronze vessels, including the writing on the Museum's *jia*: the calligraphy is fine and delicate, written in uniform lines, and in an orderly and systematic fashion.[15] Based on calligraphic style, Robert Bagley has made a comparison of the handwriting of the Saint Louis inscription with the long late Shang inscriptions on a *jiao* in the Freer Gallery of Art and a *you* in the Hakutsuru Bijutsukan, Kobe.[16]

In the Saint Louis inscription, there were further aesthetic decisions made in its writing. The scribe was particularly thoughtful about composing the two lines, especially because the text was comprised of the uneven number of twenty-seven characters which were all to fit into the limited space provided under the handle. Varying the number and size of his graphs, the calligrapher allowed for both lines to begin with the starlike cyclical characters *gui*, and then, breaking the vertical reading of the inscription, arranged the clan signs *Ya Yi* horizontally at the bottom.

The most striking aspect of the Saint Louis *jia* is its formal beauty. The decor of the vessel is elegantly restrained, done in a highly refined, nearly miniaturistic intaglio line and relievo ribbon in the upper reaches of the body. Below, the parallel bowstrings, which give further definition to the noble arches created by the legs, are

Jia No. 20 (221:1950)

produced in relief, rhythmically curving around the body. Even the animal-head handle is of modest scale, complementing the ornamental registers with its fine modelling and detail. The extraordinary pneumatic shape is based on archaic models in both ceramic and bronze. It is a formal and stylistic evocation of the past.[17]

This bronze is ascribed to the first half of the 11th century, sometime during the reigns of the kings Di Yi and Di Xin, the last two rulers of the Shang dynasty. Through his studies of oracle bones and Shang history, Xu Jinxiung determined that the Saint Louis *jia* was created during the reign of King Di Xin, the last Shang ruler,[18] who reigned for about thirty years.[19]

Selected Bibliography

Duanfang 1908, 5:32; Shang 1935; Luo 1937, 13:53b, fig. 6; Umehara 1947, 39–40; "Additional Bronzes" 1951, 63; *Handbook* 1953, 251; ; Xu Jinxiung 1968, 101; Somogyi 1986, 4–12.

Notes

1. The Saint Louis inscription has been translated in different ways by Kidder 1956, 66–68, especially 68, pl. 17; Chêng 1957; Zhang Guangyuan in a communication of formal notes to the Museum dated 31 October 1977; Bagley 1987, 525, cat. no. 103.

2. *Guisi* is the thirtieth day in the Chinese sixty-day calendrical cycle.

3. The king is one of two rulers of the late Shang dynasty; this ruler was the last (or very nearly the last) of the Shang dynastic line before the Zhou conquest and could be either King Di Xin or King Di Yi. The combined reigns of the two kings were approximately forty years in length (Chen Mengjia 1956, 210).

4. "Xiaochen" is a high official court title which is usually followed by an individual's name. According to Shang inscriptions, the position of Xiaochen was associated with officials who were court emissaries, military commanders, royal stable masters, and diviners (Chen Mengjia 1956, 505).

5. Yi is the name of a court official. Names such as Yi which follow the title of Xiaochen may also be the names of clan leaders (Chen Mengjia 1956, 507); that is to say, Yi was the head of his clan.

6. Cowry shells were the currency used by the king to reward his subjects for their service to him. The shells were in strings of ten (*peng*); thus, the total number of shells given to Yi by the king was one hundred (Chen Mengjia 1956, 557).

7. Mother Gui is the ancestor of Yi.

8. *Si* was a major ritual cycle within each king's reign, which in the late Shang and early Zhou may have come to mean and roughly correspond to a calendar year of between 360 and 370 solar days (Chen Mengjia 1956, 385–92; Xu Jinxiung 1968, 74–79; Keightley 1978, 116 n. 107).

9. *Yongri* is the day on which ritual was celebrated by drums beaten during the daytime. *Yong* was one of five minor rituals of the late Shang calendar: *Yi, Ji, Zai, Xie,* and *Yong.* Together, the rituals and their ten-day cycles comprised a year of 360 to 370 days. The drums of the *Yong* ritual are thought to have been played at specific times over three consecutive days. *Yongri* may refer to the occasion when drums were struck during the day on the second day. The *Yong* ritual cycle was composed of twelve ten-days. Because the ritual days were regular and sequential, they are used in oracle-bone and bronze inscriptions to determine dates. As important features of separate ceremonial cycles, the music of *Yong* and the dance of the *Yi* rituals are thought to comprise a category of sacrifice which is distinct from the meat, wine, and grain offerings of the other three rituals, *Ji, Zai,* and *Xie.* For a discussion of the *Yong* days see Xu Jinxiung 1968, 55–57, 118–19.

10. According to Zhang Guangyuan, the *ya* graph represents the ground plan of the four-chambered ancestral temple. The Shang aristocracy and high court officials used the character and its association with the ancestral temple to indicate their rank and power (communicated to the Museum by formal notes by Zhang Guangyuan on 31 October 1977). Ya or *yaxing* (the *ya*-shape) used in Shang inscriptions appears either alone or in association with a name. When used in conjunction with another graph, *ya* may appear before or after it or enclose it and other characters; but in every case *ya* modifies the

meaning of the other graphs. For a discussion on the meaning of the *yaxing* see Cao 1980; for examples see Barnard and Cheung 1978, 53–56.

11. *Yi* is a clan sign and may represent the position of court counselor. Composed of a pictograph of a man with a staff or scepter turning his head aside, the character resembles the word *yi*, meaning "to doubt." In the Shang system, the official known as the Yi informed and advised the king on state matters. The role of the Yi is described in formal notes to the museum by Zhang Guanyuan, dated 31 October 1977; see also Zhou, et al. 1977, 321–23, no. 2113. For an early Western Zhou example in which the graph *ya* (n. 10, above) encloses the graph *yi*, compare the inscription on the *pan* (M54:28) found at Liulihe, Beijing (*Kaogu* 1974[5]:314, fig. 22.1).

12. Chen Mengjia 1956, 233–34. For an edited but extensive version of the list and an extended annotated commentary on these inscribed bronzes, see Bagley 1987, 525–31, 535 nn. 20–22, cat. no. 103.

13. Bagley 1987, 525, cat. no. 525.

14. For a female ancestor, "worship was scheduled on the same cyclical day as [that in] her title" (Xu Jinxiung 1968, 4 and 44).

15. Ibid., 4.

16. Bagley 1987, 525.

17. Somogyi 1986, 4–12. For a discussion of archaism in later Shang bronzes see Thorp 1988, especially 63–65.

18. Xu Jinxiung 1968, 101.

19. Keightley 1978, 174 n. 20, 228, table 38. If the Saint Louis *jia* was created in King Di Xin's reign, as Xu Jinxiung maintains, then the date of the vessel can be roughly calculated. Using Keightley's table 38, the sixth *si* (year) of King Di Xin would be either 1075 or 1065 BC.

No. 21

Zhi

11th century BC
Late Shang
Inscribed: *X*
Height: 17.7 cm
Bequest of Miss Leona J. Beckmann
30:1985

The domed cover of the *zhi* is surmounted by a nippled boss decorated with individual curls. The lid fits neatly into the gently flared mouth rim and sits true to the finished lip of the vessel. The body constricts at the neck and swells at the waist, the whole supported by a flared foot. The cover, body, and foot are decorated with three narrow distinct bands of designs that are punctuated by simple elements of low relief. The lid bears a design of eye and diagonals. The eyes are composed of a boss with a short recessed line for a pupil; the diagonal divisions are filled with graduated spirals and triangles. At the neck and foot, four quilled zoomorphs in confronting pairs with heads contorted over their backs and large curled tails are divided by very low flanges. With the exceptions of the bosses and vestigial flanges, the designs are flush to the surface of the vessel and rendered in intaglio.

The relief inscription, which is a single graph cast on the interior of the lid, is designed as a square struck in the center with an X-shape with small triangles filling the interstices. A pair of low, flattened arcs brace the sides of the figure and two lines of even length extend from the square at top and bottom. A single line is evenly spaced between these linear extensions at top and bottom. The graph on this *zhi* is similar to the pictograph cast in intaglio on the interior wall of *fangding* No. 15 (108:1954). Intaglio and relievo forms of the inscription are known.[1] The reading and meaning of the inscription is uncertain. The graph has been identified with the character *lu* or "shield";[2] in its simplest form, the graph has been related to the character *zhu*, meaning "to store or hoard."[3]

Notes

1. The best-known example of the graph in relief is cast on the exterior wall beneath the handles as well as inside the cover of the late Shang lobed quadrapod *he* in the British Museum, which was formerly in the Seligman collection (Rawson 1987, 8, 94, cat. no. 8; see also Hansford 1957, 44, cat. no. A4, pl. 3).

2. Guo Moruo 1952, 215–19. The identification of the graph with the term "shield" is disavowed by Zhou, et al. 1977, 802, no. 2356.

3. Bagley 1987, 167, 169 nn. 2–3, cat. no. 8; see also Zhang Jiyun 1973, 3840, no. 7201.

No. 22

Jue

11th century BC
late Shang–early Western Zhou
dynasty
Inscribed: *Ge Xin Zu*
Height: 21 cm
Bequest of Miss Leona J. Beckmann
27:1985

Nippled bosses with intaglio curls surmount two elongated squared stems that rise from opposing sides of the interior walls to extend flush with the mouth rim on the exterior of the vessel. The deep channel spout is balanced by the prominent upswept point of the tail at the opposite side of the mouth rim. The end of the spout is finished and recedes towards the sharp edge of the mouth. The egg-shaped body is deep and rounds out in an ovoid base that is elliptical in cross section. The vessel is supported by three curving legs descending from the base and tapering to points; while each leg is slightly ridged at its outer face, it is still roughly triangular in section, with its sharpest edge facing inward toward the other two legs.

The body is fully decorated in intaglio flush to the surface starting from the neat plain border of varying width at the mouth and spout to just above the tops of the legs. The spout is ornamented with two serpentine figures that have curling tails and small round dimpled bosses for eyes with recessed pupils. Two large *taotie* dominate the sides of the vessel; below a row of vertical quills, their pupilled eyes of low oval bosses are topped by recumbent C-curls for horns. Four smaller *taotie* fill the lower portion of the design and, like the larger masks above, bear a row of quills. The plain closed handle is surmounted by a horned bovine head; another bovine counterpart is on the other side of the vessel in low relief.

The patina is a reddish dark brown which is overlaid with a thin olive-green layer. A granular corrosion of pale blue-green on the body contrasts with darker malachite on the handle. A light-beige earth fills the intaglio design.

An intaglio inscription of three characters is cast on the exterior wall of the vessel beneath the handle. The graphs read *Ge Xin Zu*, "Ancestor Xin of the Ge [clan]."

With compacted designs flush to the surface and featuring the distinctive multiple quills of earlier ornamental schemes, the decor of this *jue* represents a notable revival of Loehr Style III.[1] While the dense composition and contiguous motifs are not particular to this bronze,[2] the specific design, including large and small *taotie* and quills, occurs on only a very small number of other *jue*.[3] In 1982, a rare excavated example was discovered at Anyang dating to the late Shang; it is comparable in nearly every way to the Saint Louis *jue*, including the exact height of 21 cm.[4]

Notes

1. For a recent study on the development of the *jue* vessel type see Du 1994, 263–98, fig. 9.3.

2. The compressed and contiguous design of the Saint Louis *jue* and other *jue* are also found on *jiao*. Compare the *jiao* in the Freer Gallery (Pope, et al. 1967, 1:150–51, pl. 26). The relationship between the two vessel shapes and especially the specific design of the Museum's *jue* is strengthened by the occurrence of animal heads in relief on the face opposite the handles on both types of vessels, a phenomenon that occurs on *jiao* (but not on *jue* with designs different from the Saint Louis *jue*).

3. See Rawson 1990, 636–43, cat. nos. 104–6. A nearly identical design is on an inscribed *jue* formerly in the Seligman collection (Hansford 1957, 45–48, cat. no. A5, pl. 4). Other vessels comparable to the Saint Louis *jue* include: (a) a *jue* in a private collection in Japan (Umehara 1959–64, 3: pl. 224); (b) a *jue* in the Royal Ontario Museum, reproduced in Umehara 1964, 12, cat. no. 76, pl. 76; (c) a *jue* illustrated in Karlgren 1936, pl. 9A; and (d) a *jue* of early Western Zhou type in the Sackler collection, reproduced in Rawson 1990, 642–43, cat. no. 106.

4. The *jue* (82M1:23), inscribed *Geng Shih Fu Yi*, was excavated from Xiaotun and is illustrated in *Yinxu* 1985, 481, pls. 237–38, fig. 86, no. 4, where it is erroneously attributed to Yinxu period II. This *jue* was recovered from the same tomb that yielded the small late Shang *fangding* (82M1:44), the closest comparable vessel to *fangding* No. 16 (222:1950). Because the Xiaotun *fangding* dates to Yinxu period IV, the *jue* probably dates to the same period.

Zhi

11th century BC
late Shang dynasty
Inscribed: *Long*
Height: 14.72 cm
Mouth diameters: 9.15 cm; 7.88 cm
Foot diameters: 8.25 cm; 7.13 cm
Gift of J. Lionberger Davis
215:1950

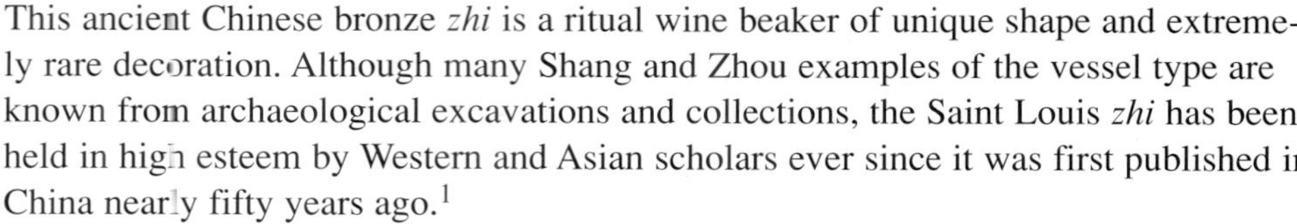

This ancient Chinese bronze *zhi* is a ritual wine beaker of unique shape and extremely rare decoration. Although many Shang and Zhou examples of the vessel type are known from archaeological excavations and collections, the Saint Louis *zhi* has been held in high esteem by Western and Asian scholars ever since it was first published in China nearly fifty years ago.[1]

The bronze beaker is not only distinguished for its reserved character and fine proportions, but also noteworthy for its shape and motifs. The body of the *zhi* is oval in cross section and rises from a nearly cylindrical base. The profile is sedate and elegant, undulating gently out from the foot to curve inward and then out again to the indented mouth.

Two distinct decorative schemes appear on the body and foot in low relief against a ground of fine spirals and whorls. First, a highly unusual decorative scheme dominates the body of the *zhi*. Four low ribs, each adorned by five to six groups of chevrons, evenly divide the ornament into four curved vertical panels. Each panel displays the silhouette of a peacocklike bird with scaly plumage and a single claw, a motif of the type known as *xiao niaowen* or "small bird."[2] Above the tail of the bird is a vertical horn-shaped motif overlaid by scales and chevrons, a design considered to be either an extension of the bird's tail or a silkworm. The genius of early bronze makers is revealed in the ambiguous character of the decorative relationship between the birds and a larger mask design: when two panels are viewed on the wide axis, the combined features of the two addorsed birds create a *taotie* with double crests, two eyes, a snout, and large subsidiary claws.

The foot design is a horizontal band divided into four sections, each bearing a pair of abstracted *kui* dragons, one of which is inverted and surmounts the other. The surface patina of the bronze is thin and of a slightly mottled gray-green color with traces of red pigment embedded in the designs of the ground. Distributed within the interior of the vessel are heavy incrustations of malachite and cuprite.[3] At the bottom of the foot is a large *long* dragon cast in intaglio, a curling serpent figure of thin lines that occupies nearly the entire area. The decoration of the *zhi* is completed along the interior curve of the footring by a small tufted dragon, coiled and grinning beneath a rolled-up nose and bearing a two-clawed foot.

The dragon at the base of the vessel forms the inscription. The image of a footed serpent was used as a pictograph of the *long* dragon, a mythical animal that is identified as a clan sign. During the late Shang and early Western Zhou dynasties, one of the main forms of inscription was the *zu hui* or "clan emblem," many of which were animal pictographs cast in intaglio.[4]

This *zhi* has long been thought to have come from Anyang, the area of the late Shang capital.[5] A 1989 Chinese report on forged bronzes sheds some light on the possible provenance of the Museum's bronze. The article states that a bronze in the Palace Museum, Beijing, the *Zi fu zhi*, is an excellent modern forgery of a Shang bronze discovered at Anyang.[6] A comparison of the Palace Museum copy and the Saint Louis beaker shows clearly that the two vessels are identical except for their inscriptions. Barring a third bronze vessel of precisely the same design, it is probable that the Saint Louis *zhi* was the model for the forgery in Beijing.[7] Moreover, according to the report, the Saint Louis *zhi* was unearthed at Anyang; this is a rare instance where a copied forgery has given a clue to the provenance of an original object.

Late Shang and early Western Zhou dates have been given to the Museum's bronze by different scholars on various grounds.[8] Specialists at the Freer Gallery

believed that its metallurgical composition (76% Cu, 16% Sn, and 5% Pb) is compatible with early Zhou bronzes.[9] The lack of consensus on the date of the Saint Louis *zhi* is further complicated first, by the highly unusual decor and unique shape of the vessel, and second, by the stylistic similarities between bronzes of the late Shang and early Western Zhou dynasties.

There is only one reliable bronze that bears a design resembling the bird and *taotie* decor of the Museum's *zhi*. The object, a ritual *gu* wine vessel of the late Shang, was published in a 1969 Christie's auction catalogue.[10] The birds and animal masks appear at the waist as well as at the lower portion of the bronze. The existence of the *gu* not only supports the late Shang dynasty date for this bronze but also marks the extreme rarity of the bird and animal mask design in ancient ritual vessels.

Selected Bibliography
Huang 1942, 2:2; "Chinese Art" 1951, 76, fig. 19; Hoopes 1951, fig 7; *Handbook* 1953, 250; Kidder 1956, pl. 4; Umehara 1964, pl. 80; Loehr 1968, 106–7; Higuchi 1973, 332; *Handbook* 1975, 270; Hayashi 1984, 2:342, no. 61; Higuchi and Enjōji 1984, no. 60; Bagley 1987, 301–2, cat. no. 49; *Haiwai* 1988, 95.

Notes
1. Huang 1942, 2:2.

2. Chen and Zhang 1984, especially 273, 281.

3. X-radiography reveals only minor repair of the rim of the vessel.

4. According to Yang Xiaoneng (private communication), the graph *fu*, for "father," appears at the left lower side of the dragon. Taken together, the *long fu* inscription might be understood as either "father of the dragon [clan]" or "the dragon [clan made this vessel for] father."

5. According to Huang Jun and Umehara Sueji, the Saint Louis *zhi* is from Anyang, but unfortunately they do not provide any supporting information. See Huang 1942, 2:2; and Umehara 1964, pl. 80.

6. Cheng, et al. 1989.

7. Max Loehr stated that "The only comparable beaker to have come to the writer's attention was in the Peking [Beijing] art market early in 1945; its whereabouts are unknown" (Loehr 1968, 106–7). There are several possibilities regarding Professor Loehr's sighting of a similar *zhi*: (1) the "comparable beaker" may be the forgery now in the collection of the Beijing Palace Museum; (2) there is a third and unknown *zhi* of similar design; or (3) Loehr may have indeed seen the Saint Louis *zhi*. Furthermore, a *zhi* of similar shape and design was illustrated by Kunsthaus Lempertz in their auction catalogue *Ostasiatische Kunst* 1992, 17, lot no. 35.

8. Kidder concluded that the *zhi* is a product of the Shang dynasty (Kidder 1956, 38). Hayashi Minao dated it to late Shang (Hayashi 1984, 2:342, no. 61). Loehr assigned the vessel to Style V of his Shang stylistic sequence (Loehr 1968, 106). Higuchi and Enjōji assigned it to early Western Zhou (Higuchi and Enjōji 1984, no. 60). On the basis of the bird motif, the vessel is dated to the Western Zhou circa 1050 BC in Moffitt and Voris 1988.

9. From a letter dated 21 September 1971 by W.T. Chase, Head Conservator, Freer Gallery of Art/Sackler Gallery.

10. Christie's *Fine Chinese Ceramics* 1969, lot 173, pl. 25. There are extremely few bronze designs to compare with the bird and animal mask design of the Saint Louis *zhi*. Hayashi Minao cites a late Shang dynasty ritual bronze *jue* with an interesting variant of the combination bird and animal mask. (Hayashi 1984, 1:58–59). For further comments on the *gu* and the Saint Louis *zhi*, see Bagley 1987, 301–2, cat. no. 49.

No. 24

Fanglei

late 11th century BC
early Western Zhou dynasty
Inscribed: *X*
Height 62.7 cm
Width: 37.8 cm
Museum Purchase 2:1941

The *fanglei* or squared *lei* appears around the late 13th century BC among bronzes excavated from the Shang capital of Anyang.[1] The basic features of the *fanglei* appear well developed and included a pyramidal cover and knob, three lug rings, squared shape, animal heads at the shoulder, bowstrings at the neck, medial ridge dividers between motifs, and animal heads with pronged horns.[2] But during the late Shang and early Western Zhou, the older, more conservative model was wonderfully transformed into visually intriguing vessels with bristling, rather aggressive appendages and shapes.

The cover of the Museum's *fanglei* is a four-sided pyramidal shape surmounted by a miniature pyramid on a square stem. The knob bears an intaglio everted *taotie* with fangs. The neck is square in cross section and joins a four-sided but swelling rounded shoulder that slopes and gradually narrows down along the body to a splayed raised foot. The decorative order of the *fanglei* is based on an interplay of flat plain surfaces represented by the long-nosed creatures below the shoulder and the scooped and multi-layered elements of the double *taotie* on each side. These in turn contrast with the intaglio-lined and rounded surfaces of the small dragons in very low relief. The flanges on the lid, neck, shoulder, and foot have hooks and spurs at one terminus but not at the other; the length of each flange bears parallel intaglio lines interspersed with spurred curls. The neck is decorated with hook-nosed dragons with spur feet, C-curl horns and tails, and a large lower jaw. The shoulders present dragons in profile with long snouts and curled lower jaws. They have curving horns and tails as well as hind legs with dewclaws. The body hosts small dragons with spurred proboscides, large eyes, and C-curl horns. The undershot jaw juts from under the eye and its legs are hooks and spurs. The foot is ornamented with small dragons whose hooked snouts rise above large lower jaws. The raised and curling tail bifurcates from the lower body and its two spur legs. On the body are large *taotie* with dished C-curl horns and hooked plumes. To the side of large oval eyes with recessed pupils and narrow brows are pointed ears of dished leaf shapes with single hooked quills in intaglio. The effect of a nasal crest and snout is made more pronounced by the spurred and hooked flange.

The body of each side bears a large mask under which is a smaller mask of similar design. The vessel has three lug handles, each ornamented by animal heads with scaled necks and bearing antlers in high relief. Snub-nosed and lantern-jawed, the heads teem with files of even, sharp teeth and round bulging eyes and protruding ears marked by a cross of lines that suggest the outer convolutions of the aural canal. The spiky character of the paired span of saw-toothed antlers is reinforced by the ornaments of inverted curvilinear V-shapes. The two lugs at the shoulders enclose large flat ring handles with a design of eyed diagonals. On the exterior wall of the vessel, beneath the two lugs, are cicadas in relief. The third lug is placed on the face of the body, bisecting the double *taotie*. Underneath the third lug is an elementary fret design in relief. With the exception of the vessel's shoulder which bears animal heads or ornamented lugs, the vertical medials of every decorated register are marked by thick high flanges that have hooks and spurs.

The vessel is inscribed with a puzzling and unusual graph on the interior walls of the neck and cover. The graph is a large character composed of four vertical lines, two long parallel lines flanking two rows of three short lines each. Despite many attempts by scholars to decipher it, the meaning remains uncertain, but most agree that it is a name.[3]

The most intriguing theory identifies the inscription as an oracular sign. This graph and very similar ones have been recorded on five other early Western Zhou bronzes. A study of Shang and Zhou divinatory inscriptions argues that such graphs are symbols similar to the diagrams found in the *Tai Xuan Jing* system of divination.[4] Compiled in the Western Han dynasty by Yang Xiung (BC 53–18 AD), the *Tai Xuan Jing* was similar and perhaps contemporary to the Zhou *Yi Jing* (I Ching, the Book of Changes). Both the *Tai Xuan Jing* and the *Yi Jing* were ancient forms of divination that used the thin straight stalks of the milfoil plant to consult the oracles.[5] The diagrams, which are determined by counting milfoil stalks, were different combinations of three, four, or six whole or broken lines, horizontally aligned. The Saint Louis inscription resembles the diagrams to a remarkable degree.[6]

Another theory proposes that divinatory graphs are probably clan names adopted by officials who received lands from the king.[7] Upon claiming the royal grant, the official performed prognosticatory rites with milfoil stalks. The diagram which resulted from the divination was then given as the oracular name of the place. The new name was, in turn, used as a clan name or emblem.[8] Such names, it may be deduced, were inscribed on bronzes in exaltation of the official, his clan, and royal favor.

By the late Shang and early Western Zhou, elaborate and flamboyant forms of the *fanglei* and other vessel types were created with prominent hooks, flanges, and in some cases long curving beams.[9] The resemblance of the Saint Louis *fanglei* to this visually provocative late style has prompted scholars to speculate on its date and provenance. In 1940, James Menzies believed the vessel was Shang and from Anyang.[10] Much later, in 1985, the *fanglei* was linked to two other early Western Zhou bronzes, a *fangyi* in the Harvard University Art Museums, and a *ding* in the Asian Art Museum of San Francisco (Brundage Collection),[11] both of which are said to have been found in 1929 at Baoji Xian in Shaanxi province. Based on their similarities of motifs, scale, casting, and flanges, the three bronzes were thought to comprise a distinct, albeit small, set.[12] In essence, the Saint Louis *fanglei*, the Harvard *fangyi*, and the Brundage *ding* would comprise a "third" Baoji set.

The Baoji altar sets are among the most famous of Chinese bronzes. The sets take their name from Baoji Xian, a county in Shaanxi province that since the early 20th century has been associated with the discovery of two bronze caches. The first Baoji set, also known as the Duanfang set, consists of thirteen wine vessels, six ladles, and a short bronze table; the bronzes reportedly were excavated in 1901 at Doujitai, Baoji Xian, Shaanxi province. Between 1902 and 1903, the set was acquired by the collector and connoisseur Duanfang (1861–1911), who noted it in his catalogue on inscribed bronzes, *Taozhai jijin lu* of 1908. The twenty bronzes were later purchased by the Metropolitan Museum of Art, New York in 1924.[13] The second Baoji set consisted of nine wine vessels, two food containers, and a long bronze table, said to have been excavated from Daijiagou, also at Doujitai in Baoji Xian. The second set of vessels has been dispersed to various collections.[14]

The integrity of the two Baoji sets, on both ritualistic and stylistic grounds, has been questioned and much about both sets suggest that they may be later assemblages.[15] Furthermore, the indigenous bronze style of the early Western Zhou Shaanxi province, the reputed source of the Baoji sets, represented thus far by the Yu Bo Ji and Yu Bo excavations at Baoji, is related to but distinct from the Baoji sets, including the proposed group of three bronzes from Harvard, Brundage, and Saint Louis.[16] The Harvard and Brundage bronzes, which bear distinctive ribbed panels and bands, are certainly related to a number of early Western Zhou vessels with similar decoration; the Saint Louis *fanglei*, however, is of a different decorative order.

Notes

1. See the *lei* (76AXTM5:856) from Fu Hao Tomb 5 at Xiaotun in *Henan chutu* 1981, 26, cat. no. 150; 126, pl. 150.

2. For a discussion of the development of the *fanglei,* see Bagley 1987, 28–29.

3. Several interpretations have been proposed over the last five decades. James Menzies deciphered this graph in 1940 as the character for "rain," *yu*, which may have referred to an official name or title (*An Exhibition* 1940, cat. no. 20, pl. 4). Menzies later revised the reading *yu* (rain) to *chuan* (river), in a private letter dated 20 January 1941 to the Shanghai dealer C.T. Loo who owned the *fanglei* which was later purchased by The Saint Louis Art Museum, then known as the City Art Museum of St. Louis. J. Edward Kidder assumed the character to be a name and to represent "water flowing between two banks" (Kidder 1956, 71). The graph has also been read as *chao,* meaning "tide" (Zhang and Liu 1981, 163) or a name meaning "a damp place." Jessica Rawson notes that according to one opinion the Saint Louis graph may read *ce,* meaning "documents" (Rawson 1990, 246, 247 n. 7).

4. The symbols are similar to the *bagua* (the Eight Diagrams, also known as Eight Trigrams) of the *Yi Jing* (I Ching) system of divination. See Zhang and Liu 1981, 157, 160, 163, fig. 2.26–29, graph 1.26–29; see also the English translation by Edward L. Shaughnessy in Zhang and Liu 1981–82, especially 53–54.

5. The Eurasian herb *Achillea millefolium*, a strongly scented perennial with straight stalks and medicinal properties, known also as the yarrow plant. In divination, the stalks were typically kept in a bamboo tube; later, milfoil stalks were substituted with thin slips of bamboo.

6. The graph on the Saint Louis *fanglei* corresponds most closely to the *Tai Xuan Jing* diagram *zeng*, meaning "to increase." Zhang and Liu show the character *zheng* (Zhang and Liu 1981, 163). However, as shown in the *Tai Xuan Jing*, the diagram is read as *zeng* which is composed of two whole parallel lines, one above and one below a pair of horizontal rows of two broken lines each. See Yang 1965–66, 1:15a–16a.

7. Zhang Zhenglang 1980; this article was subsequently published in a collaborative translation. See Zhang Zhenglang 1980–81.

8. Zhang Zhenglang 1980–81, 90–92. This can be seen as the ancient Chinese equivalent of the English practice of calling lords by their lands: York, Norfolk, or Lancaster.

9. For a discussion of the flamboyant bronzes, see Rawson 1990, 35–57.

10. In 1940, James M. Menzies erroneously dated the bronze as Shang and misattributed the provenance of the vessel to Anyang (*An Exhibition* 1940, cat. no. 20, pl. 4). In 1937, J. Leroy Davidson recognized that the Museum's *fanglei* was of early Western Zhou date, further noting that it was "of a slightly later date" than the Duanfang (first Baoji) altar set at the Metropolitan Museum of Art (Davidson 1937, 32).

11. For a reproduction of the Harvard *fanglei* see Mortimer 1986, 17, cat. no. 7; for an illustration of the Brundage *ding* (B62 B144) see d'Argencé 1977, 72–73, cat. no. 28.

12. Murray 1985, 44.

13. For an illustration of the first Baoji set see Watt 1990, 18–19, pl. 17.

14. For an illustration of the second Baoji set, see Bagley 1980, 207, cat. no. 48; and Rawson 1990, 156, figure A.

15. For summaries of the histories of the Baoji sets and questions concerning their relationships, see Bagley 1980, 207, cat. no. 48; Rawson 1984, 28 n. 4; Rawson 1990, 155–60.

16. Rawson 1990, 158–60; *Wenwu* 1983(2):1–11; *Wenwu* 1988(3):20– 27.

No. 25

You

11th–10th century BC
early Western Zhou dynasty
Inscribed: *Shi Po Zhi zuo Fu Gui yi*
Height: 26.7 cm
Gift of J. Lionberger Davis
225:1950

Since its discovery more than two hundred years ago, this *you* has continued to hold the scholarly interest of collectors and connoisseurs who avidly recorded its inscription in their voluminous and detailed catalogues. Among the very first to take note of this bronze was the great Qing scholar Ruan Yuan (1764–1849).

Shortly after it was discovered in Shandong province in 1791, the bronze caught the attention of Ruan Yuan, who had just completed cataloguing the imperial calligraphy and painting collection in the Forbidden City. Upon assuming his post as director of education in Shandong in 1793, Ruan Yuan began work on the *Shanzuo jinshi zhi*,[1] a study of inscriptions on stone and bronze where he recorded the unearthing of the present *you* at a place called Liushanzhai in the county of Linqu.[2] In ancient times, during the Spring and Autumn period, Linqu was called Quyi in the State of Qi.[3] By linking the names mentioned in the *you*'s inscription with historical texts and important clans, Ruan Yuan found evidence that the Museum's bronze was a ritual vessel of great antiquity.[4] For the next generation or so, the Saint Louis *you* circulated among connoisseurs until it was acquired by the distinquished collector of bronzes, Chen Jieqi (1813–84; *jinshi* 1845), and recorded in his *Fu Zhai jijin lu*.[5] The bronze has enjoyed similar interest ever since.

The cover of this *you* is a broad low dome that gradually slopes to meet a high concave collar; the deeply recessed lines and granulation of the stemmed knob as well as the sharp pronounced edge are distinctive features of the lid. A bail handle broadly arches over the vessel along its long axis and joins two opposing lug rings that protrude just below the lip of the vessel. The container's body swells, then constricts to a high concave and flared foot with a molded base. The bail handle is a flat strap ending in rings; its rounded crown is decorated in low relief with hooked and spurred curls and C-shapes.

Extending from posts at the ends of the handle are two sculpted animal heads with capped bottle horns, short snouts, and bulging eyes. The cover is decorated with four panels, each composed of a bird with plumed crest, sharp curved beak, taloned feet, and spurred elongated tail feather. The birds are joined by a contorted animal with antlers, a hooked snout, spurred body and a long curling tail. This same design decorates the band on the body; each panel is bound by the lug rings and a small *taotie* in high relief and a thin bowstring border. Bowstrings surround the foot and are also featured as borders and ornaments for the lid.

The patina is a generally mottled surface of reddish brown and light malachite green color with spidery patches of black tin corrosion. Areas of the vessel are overlaid with thin crusts of granular malachite and underblooms of bright cuprite.

The vessel is inscribed with seven characters in two columns on the interiors of the lid and body.[6] The inscription, *Shi Po Zhi zuo Fu Gui yi*, may be read "Shi Po Zhi made for Father Gui this vessel."

The Saint Louis *you* can be favorably compared to several very similar vessels in Western and Japanese collections. Each is distinguished by the sectioned finial or knob on the lid and the smooth rounded forms of its animal decor. Their body shapes also conform to the bulging oval cross section with a slight attenuation of the upper portion, an effect that is enhanced by the plain high collar of the lid.[7]

Selected Bibliography
Luo 1937, 13:26b; Kidder 1956, 64–66, pl. 16; Barnard and Cheung 1978, 557, no. 624

Notes
1. Hummel 1943, 339.

2. The inscription of this *you* was recorded by Ruan Yuan in the *Shanzuo jinshi zhi* (1797) while he was director of education in Shandong under the supervision of Bi Yuan (1730–97). The entry describes the *you* as being excavated at Liushanzhai, Linqu Xian, Shandong in 1791 and then sold.

3. Chen Zhengxiang 1975, 1291; and Zhang Jiyun 1973, 11881–82.

4. Ruan Yuan believed the *you* to be of the Spring and Autumn period (Ruan 1797).

5. Chen Jieqi recorded the inscription in his *Fuzhai jijin lu* (Chen Jieqi 1918, 2:5). According to Thomas Lawton, the Saint Louis *you* was formerly owned by Chen Jieqi, a Shandong collector of bronzes. Chen was a younger contemporary of the noted Shandong collector of rubbings Wu Shifen (1796–1856) who also recorded the *you*'s inscription (Wu Shifen 1895, 2:1, fig. 9). Thanks and appreciation to Lawton for the identification of the Saint Louis bronze with the Ruan Yuan and Chen Jieqi catalogues (conveyed in a letter dated 12 January 1990).

6. The character *yi*, or "vessel" is in the first column of script on the vessel body; on the lid, the character appears in the second column.

7. See Karlgren 1952, 54, cat. no. 17, pl. 24; the Victoria and Albert Museum piece in Watson 1962, pl. 16; and examples of the same early Western Zhou vessel form and design treatment on the *you* in the Idemitsu Bijutsukan and Hakutsuru Bijutsukan illustrated in Hayashi 1984, 2:271, nos. 121–22.

You No. 25 (225:1950)

No. 26

Gui

11th–10th century BC
early Western Zhou dynasty
Height: 14.6 cm
Mouth diameter: 18.3 cm
Museum Purchase
203:1916

The spectacular coiled dragon of the Saint Louis *gui* is completely unknown in Shang designs.[1] As a singular early Western Zhou motif, it is associated with excavated bronzes of Shaanxi province[2] but has also been found at archaeological sites in western and northeastern China.[3] In early Western Zhou bronzes, the highly animate, even kinetic, dragon is often coupled with high hooked flanges to produce visually provocative and fantastic vessels in the forms of *gui*, *zun*, *you*, and *lei*. Some *gui* were integrally cast with high square bases to further enhance the bronzes' eccentric and fabulous character.

The reserved shape of the bowl of this high-footed *gui* is offset by its sensitively rendered animal-headed handles, pointed hooked flanges, and sharply molded base as well as the plastic treatment of the distinctive dragons around the body. The decoration is rendered in relief of differing levels against a *leiwen* ground. At the handles, sculpted animal heads, which have leaf-shaped ears that lie back to the beveled everted lip of the vessel, bird bodies with low-relief wings and sweeping spurred lines in intaglio. Below are hooked tails and taloned claws raised against the flat surfaces of pendant rectangles.

The two pairs of confronting dragons encircling the *gui* are shown in profile, their prominent and trumpeting trunks curving above opened jaws and curling tongues. A single curled horn rises above the large boss eye and the quilled hooked vertical brow and clawed front leg. A large volute in relatively high relief forms a high spiraling tail. A medial hooked flange surmounted by a small mask separates the confronting animals. The base is decorated with four serpentine dragons bearing bottle horns, claws, and curling proboscides and tails, their sinuous bodies ornamented by a line of cowry designs. These dragons are divided by short hooked flanges.

The patina is a fine dark earthen color overlaid with a granular corrosion of dark malachite and an orange to dark red cuprite that is usually associated with a scaly lead carbonate.

The creative vigor of the early Western Zhou bronze makers, so obvious in the shapes and forms of their vessels, is also evident in the often bewildering minutiae of the *leiwen* ground. On the Museum's *gui*, the *leiwen* is wonderfully irregular. Tight miniature volutes spin in concert with the coils of the big dragons; other ground-fill takes the form of elaborate S-curls and C-curls and triangles in multiple layers. Here and there, small T-curls and C-curls with stems are encapsulated in oval devices.

The early Western Zhou coiled dragon is represented by three basic designs that are distinguished from one another by a small detail within the mouth of the beast. The jaws of the Saint Louis *gui* dragon have a broad flat curl with points that represent the upper and lower teeth totally flush to the vessel surface. In another form, the monster is shown with two small oblong bosses that project in relief for a literally more toothy presentation. In the third form, the lower jaw is curled and pointed under the beast's gaping upper jaw and raised trunk.[4]

Another small difference between vessels with the coiled dragon motif appears at the animal-head handles. Some vessels like the Museum's *gui* have bestial heads that have low profiles and flattened ears which conform to the shape of the handles. In other instances, the heads have pronounced contours enhanced by large three-dimensional horns that are often antlered with spurs and hooks.

Selected Bibliography

"A Bronze Sacrificial Vessel" 1916, 9; "A New Installation" 1917, 12.

Notes

1. For a discussion of the coiled dragon motif, see Rawson 1990, 37, 55–57.

2. In Shaanxi province, a pair of *gui* on square bases, *zun*, and *you* were excavated at Gaojiabao, Jingyang Xian in 1971 (Ge 1972, 5–8, 67) and in 1966, a *gui* and *lei* were found at Hejiacun, Jishan Xian (Zhang Shui 1972, 26, fig. 4).

3. In Western China, a *gui* was found in tomb 1 at Baicaopo, Lingtai Xian, Gansu province (*Kaogu xuebao* 1977[2]:99–130) and two covered *lei* were discovered at Zhuwajie, Peng Xian, Sichuan province in 1959 (*Wenwu* 1961[11]:5, 28–31). In the northeast, a covered *lei* was excavated in pit 2 at Beidongcun, Kezuo Xian, Liaoning province in 1973 (*Kaogu* 1974[6]:364–72, pl. 1).

4. For the other forms of the coiled dragon, see the bronzes found in 1971 at Gaojiabao, Jingyang Xian, Shaanxi province. See also the *gui* dragon with projecting teeth and the *you* and *zun* dragons with curled lower jaws in Bagley 1980, 222–23, pls. 49–51.

No. 27

Gui

late 11th–early 10th century BC
early Western Zhou dynasty
Height: 21.3 cm
Mouth diameter: 27.3 cm
Gift of J. Lionberger Davis Art Trust
123:1951

The *gui*, as a basic type in ceramic,[1] marble,[2] and bronze, is a raised footed vessel. Here the container is integrally cast to a hollow square pedestal to further enhance its height and presentation. Such vessels have their probable prototypes in the Shang dynasty[3] and were not uncommon during the early Western Zhou.[4]

The very low relief decor of this *gui* lies against a plain unornamented ground. Two large well-formed handles project from the flaring beveled lip and join as closed C-shaped lugs at the swelling portion of the bowl. The handles are surmounted by animal heads with large upswept pointed ears neatly detailed in intaglio curls, and bulging rounded forms that define the creature's eyes, cheeks, and nostrils. From their open mouths issue the headless bodies of two birds, the barbed wings of which curve down the length of the handles.

Pendant below the lugs, the birds' curled tails and claws are rendered on flat rectangles. The *taotie* on the vessel and the stand are marked by a high medial ridge that forms the nasal crest and snout. Large C-curls with intaglio lines create the horns rising over the horizontal brow and small leaf-shaped ears. The eyes, which are oval bosses with recessed pupils, stare out above the hinged upper jaws and hooked fangs; the nostrils are shaped by small paired bosses.

The masks are flanked by amorphous vertical motifs with hooks and spurs. The foot of the vessel is decorated with two pairs of confronting dragons whose sinuous elongated bodies end in curling tails. The four sides of the pedestal bear four *taotie* that are nearly identical to the masks on the vessel above; even the horns of the animals are rendered with similar intaglio hooked and spurred curls.

The patina of both the vessel and its high base is a smooth, even, silvery gray-green color. The color is darker and shiny at the handles where there are black deposits, especially at the flat pendant rectangles. The surface of the bronze bears patches of white and yellow accretions, possibly lead carbonate corrosion, beneath which bloom cuprite red and orange. Speckling the pedestal, malachite green and dark azurite blue mix with rust and orange cuprite. The piece has been repaired.

The smooth decoration of this *gui* and the Museum's other *gui* which is similarly treated (No. 28; 540:1956) may be related to a small group of earlier transitional vessels of the Shang–Western Zhou with rounded softly modelled animal-head handles and, more specifically, smooth, rounded, very low relief *taotie* designs rendered on an absolutely plain ground.[5]

In terms of its elevated presentation, proportions, and motifs, the Museum's bronze follows a model that was established by the inscribed Li *gui*, a very early Western Zhou vessel with a square base, excavated in 1976, that records the royal reward of bronze metal to the official Li just seven days after the capture of the Shang capital by the Zhou king Wu Wang.[6] Although the Saint Louis *gui* is of a slightly later date and of a different decorative order than the Li *gui*, the Museum's bronze is indicative of the popularity of the stand-and-vessel combination during the early Western Zhou and its continued use in later times.

Selected Bibliography
"Additional Bronzes" 1951, 62; *Handbook* 1953, 253; Kidder 1956, 77–78, pl. 21.

Notes

1. *Kaogu* 1977(1):20–36, pls. 5.1–2.

2. Mizuno 1958, 89, fig. 243.

3. Rawson 1990, 33–34, fig. 29; see also Bagley 1987, 35, 120, fig. 172.

4. For a discussion of the development of bronze vessels with square bases see Rawson 1990, 33–34, figs. 25–29.

5. This group includes the *gui* in Cologne (illustrated in Goepper 1972, cat. no. 10) and the Ya Chou *gui* (formerly von Lochow collection) and its inscribed cover (now in the Museum of Far Eastern Antiquities, Stockholm); for images of both the vessel and cover see Rawson 1990, 362, figs. 38.1–2. Both the Cologne and Ya Chou bronzes may date to the Western Zhou period. Bagley believes the Cologne piece is possibly early Western Zhou (Bagley 1987, 517, fig. 102.4). The cover of the Ya Chou *gui*, which is separated from the vessel, is surmounted by a high flared circular collar by which the lid can be grasped and removed. This ring is typically a Western Zhou device and unknown on Shang covered bronzes, which usually have small knobs; compare the Saint Louis *lihe* (No. 17; 287:1955).

6. The Li *gui*, which is noteworthy for its long inscription of thirty-two characters, was discovered in Lintong, Shaanxi province in 1976 along with four later ritual vessels and numerous bronze implements (*Wenwu* 1977[8]:1–7, pls. 1, 2.1–2, figs. 2, 4.2, 18.1–4). For a discussion of the Li *gui* and its inscription see Bagley 1980, 198, 203, cat. no. 41.

Gui No.27 (123:1951)

No. 28

Gui

early 10th–late 11th century BC
early Western Zhou dynasty
Inscribed: *Zuo bao zun yi*
Height: 15.2 cm
Width: 25.2 cm
Gift of Miss Leona Beckmann
in memory of her brother,
Dr. J. William Beckmann
540:1956

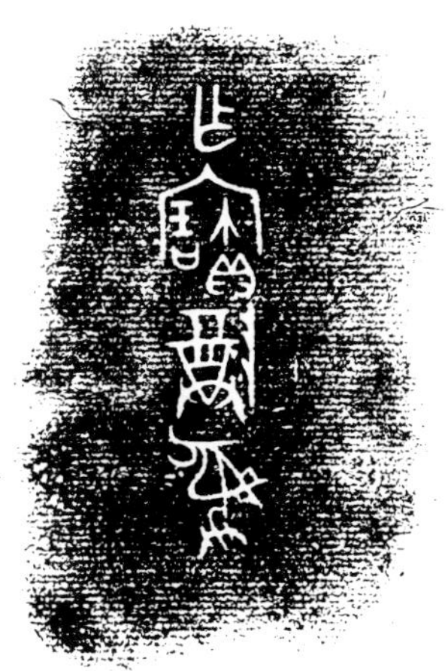

This *gui* is of a well-known early Western Zhou type characterized by smooth low-relief decor against a plain blank ground. In small but significant contrast to the flat quality of the overall design are the pairs of vividly portrayed animal heads at the handles and under the rim. Derived from bold Shang prototypes, the seemingly startled expressions of these carefully modelled heads enliven and soften the severe, stately presentation of the vessel.

The flaring lip of this bronze is flattened and beveled; no part of the decor intrudes on its plain smooth surface. The two lug handles feature short-faced animal heads with small leaf-shaped ears and capped bottle horns. The headless bird bodies that issue from the animals' mouths show low-relief elements and wings; intaglio lines on flat pendant hooks below delineate the generalized tail and claws of the bird. The *taotie* is a mask of related but dispersed elements: the curving C-shaped horns are beveled and dished and rise prominently above the face, exceeding by a fraction the upper border of the decorative band. With staring boss eyes showing recessed pupils, the masks have sharp-tipped leaf-shaped ears that are slightly dished and marked by a cross of hooked lines. A small horned animal head decorates the hooked crest and low-relief beadlike ridge. The *taotie* are flanked by anomalous hooked and spurred vertical motifs. Around the foot are four pairs of horned dragons in confrontation at the hooked medial lines marked by a low flange.

The bronze is of a fairly even light green color with a blue undertone; its surface is smooth with finely granulated patches of darker green. The animal heads and handles show a darker, shiny, and finely patinated veneer.

The intaglio designs on the animal-headed handles are noteworthy not only for their fine crisp rendering but also for introducing an altogether new decorative order of small motifs to bronze design. While the hooked and spurred intaglio curls of the Shang can still be found as tokens of the past on the flat pendant hooks below the handles, the zigzags, crosses, and daggerlike marks on the animal heads are innovations of the early Western Zhou bronze makers.

The inscription, which is cast on the interior floor of the bowl, consists of a single column of four characters and may be read *Zuo bao zun yi*, "Made this precious sacrificial vessel." The absence of the names of the person who commissioned the bronze and the ancestor for whom the bronze was made is not unusual in Western Zhou inscriptions.[1]

Selected Bibliography
Noel and Cheung 1978, 627, no. 816.

Notes
1. Following conventional formulae, the inscription might read "[X] made [for ancestor X] this precious sacrificial vessel." The cataloguers of the Chinese bronze collection at the Freer Gallery of Art commented that "Forgers have often contented themselves with limiting their efforts to this incomplete sentence...we can trace their manipulations from the period of the Sung [Song] catalogues through the early Ch'ing [Qing] catalogues up to the present. It cannot, of course, be surmised that every such inscription text is spurious" (Pope, et al. 1967, 1:378, cat. no. 67).

No. 29

Danglu

11th–10th century BC
Western Zhou dynasty
Height: 29.6 cm
Gift of J. Lionberger Davis
218:1950

The *danglu* is a decorative and protective frontlet secured to the forehead of a horse. In form and function, it serves as defensive bard or horse armor, specifically the chamfron which in its largest European examples can cover the animal's entire face. Archaeologically excavated pieces from the Western Zhou site at Baifu show three types of *danglu*,[1] all of which are related by their predominantly rounded forms and horned shapes. The Saint Louis piece, with its three-dimensional character, is indicative of the most expressive, decorative, and sculptural of the three types.[2]

The feline mask of this bronze fitting surmounts a tapering column which is sparsely decorated in intaglio with three widely spaced chevrons patterned with recessed dots. Crowning the head, a curled pair of broad pointed horns are rendered in multiple levels and decorated with a curving spurred recessed line. The horns meet the low flat ridge of the brow in large twin volutes positioned just above the eyes. These eyes, which stare from sockets sculpted between the brows, snout, and cheeks, bulge as round bosses with recessed pupils. The epicanthi are rendered in rounded lines, their catlike slant enhanced by the widely spaced, bluntly protruding temples. The overhanging brows, which are ornamented with spurred and hooked curving lines, are separated by a pair of curvilinear devices along the snout. The feline's muzzle is created from unculating layered ridges at the cheeks and is decorated with nearly parallel lines of dots; its snub nose is delineated by two curved lines ending in deeply recessed dots for nostrils. Extending below the head is the hollow column which is roughly hemispherical in cross section and closed at the nether end.

The patina is a fine, smooth, grainless surface of dark olive-green color which occurs in small scattered patches. The bronze is evenly corroded to a mottled light bluish-green. The columnar stem of the bronze has been repaired.

Selected Bibliography
"Additional Bronzes" 1951, 60; *Handbook* 1953, 255.

Notes
1. Sixteen *shi danglu* (decorated chamfron) were excavated from the wooden chambered tombs of the ancient Western Zhou state of Yan at Baifu in Changping near Beijing. The chamfrons are divided into Types I, II, and III. In a description of Type III, of which there are six examples found at the site, "the upper portion is in the form of an animal's head, the lower portion is in the form of a long shank" (*Kaogu* 1976[4]:257, nos. 5–7, especially fig. 18).

2. Pieces comparable to the Saint Louis chamfron are reproduced in *Important Chinese Works of Art* 1983, cat. no. 34; and "The Exhibition" 1934, pl. 26, figs. 2–3.

No. 30

Snake

11th–10th century
early Western Zhou dynasty
Diameter: 24.5 cm
Gift of J. Lionberger Davis Art Trust
124:1951

The coils of this rare bronze are rendered as a volute swirling about the head of the snake at its center. The body of the creature is slender, its width gradually narrowing to the point of the tail; the whole of its length is hemispherical in cross section, its edges molded in a single step. The expression of the serpent is alert, its eyes and nose both emphasized by their projection, and a pair of spiralled horns provides the snake with a penetrating look. A diamond-shaped design ornaments the head and fine intaglio lines file down the snout from the brow.

The patina is an even silvery-green color with irregular patches of pale-green malachite and a light-beige earth; flecks of cuprite and dark green are occasionally seen. The back of the bronze shows fine crossing ribs at irregular intervals along the body. Thirteen thin straps are arranged along the length of the snake, possibly to provide a means of attachment to a background as decoration.

There is a comparable coiled bronze serpent in the Museum of Fine Arts, Boston.[1] The Saint Louis bronze was reportedly found at Xicun, Luoyang, Henan.

Selected Bibliography
"Additional Bronzes" 1951, 59; *Handbook* 1953, 254; *Découverte* 1954, 103, no. 406; Kidder 1956, 80, pl. 22.

Notes
1. Museum of Fine Arts, Boston (36.254). My thanks to Tony Carter of London for confirming this information.

No. 31

Zhi

10th century BC
early Western Zhou–middle Western
Zhou dynasty
Inscribed: *Yuan zuo lüyi*
Height: 14.9 cm
Gift of J. Lionberger Davis
27:1951

This extraordinary vessel is remarkable for its sculptural form and anthropomorphic features. Even more noteworthy is that it is apparently one of a pair of *zhi* cast with nearly identical design and the same four-character inscription; the other bronze is in the Avery Brundage collection of the Asian Art Museum of San Francisco.[1]

Supported by a broad plain concave foot, the body of this *zhi* swells expansively, constricts at the neck, and then flares slightly to a thickened, finished lip. From a rounded neck, the waist bulges at four points to provide a squarer cross section. The high dome cover is surmounted by a simple solid half-ring handle.

The narrow band of decoration at the neck consists of twelve rondels against a *leiwen* background, each rondel punctuated in the center by a small rounded point surrounded by four hooks. Four prominent bosses dominate the lower body. These are the staring eyes of two *taotie* with recessed pupils. The eyes themselves are recessed into shallow pockets between the nasal ridge and puffed cheeks, giving a sculpted surface for the raised curvilinear decor. The brows are marked by lozenges and crests flanked by recumbent C-curls for horns. Nostrils, fangs, and jaws are delineated by hooks and curls with paired hooked C-curls marking the cheeks. Two *taotie* similar to those on the body decorate the cover. The handle bears four small bosses with recessed points; these are accompanied by truncated C-curls in similar low relief and marked by recessed lines. All motifs are rendered in a thin even raised line set against a *leiwen* background. Both the cover and body of the bronze have a slightly granular texture. The splotched and mottled coloring of malachite and reddish brown cuprite corrosion cover a reddish brown patina.

The bronze is cast with two inscriptions which appear on the interior floor of the body and on the interior wall of the lid. The inscriptions are composed of four characters in two columns of two characters each. The inscriptions read *Yuan zuo lüyi* and may be translated as "Yuan made this *lü* vessel." The meaning of the word *lü* is problematic and open to many interpretations.[2]

The linear use of thread relief has Shang antecedents at Anyang, among the finely engraved marble sculpture and some stone vessels where thin lines were carved in relief on a plain ground.[3] Early Western Zhou examples reveal a later but similar appreciation for the elegance of thread relief, particularly when laid against and contrasted with a fine ground of *leiwen* and the kinetic forms of the whorl and rondel.[4] Both the Saint Louis and San Francisco bronzes are comparable to a third covered *zhi* in Stockholm of similar date and decor, featuring a small oval ring-collar handle on the lid, but lacking an upper band of whorls at the neck.[5]

Selected Bibliography
Loo 1941, cat. no. 55; Hoopes 1951, fig. 9; *Handbook* 1953, 254; Kidder 1956, 72–73, pl. 19; Barnard and Cheung 1978, 640, no. 851.

Notes

1. d'Argencé 1977, 88, cat. no. 36, front right. These two vessels may in fact be the pair of bronzes exhibited by C.T. Loo at his sale in 1941–42 (Loo 1941, cat. no. 55).

2. As a modifier to the word *yi* or "vessel," the character *lü* may mean "sacrificial." However, the character *lü* has many meanings. See Liu 1902; van Heusden 1952, 98; Karlgren 1952, 56, cat. no. 18; Kidder 1956, 72; Chêng 1956, especially 14–15; Rawson 1990, 223, 371, 467, 487, cat. nos. 3, 40, 61, 66.

3. Compare the marbles excavated at Anyang and carvings in Chu 1983, color plates, figs. 3–10. See a fragment of a marble vessel in Huang 1937, 33b, no. 2 cited by Rawson 1990, 482–84, cat. no. 65.

4. See *Wenwu* 1977(12), figs. 9–10, 43, 47–48; and Pope, et al. 1967, 2:304–9, 402–5, cat. nos. 54, 73.

5. See comparable *zhi* in the Museum of Far Eastern Antiquities, Stockholm, in *Haiwai* 1988, 41.

No. 32

Gui

10th century BC
late early Western Zhou–early
middle Western Zhou dynasty
Inscribed: *Zuo Fu Yi Zi*
Height: 17.7 cm
Mouth diameter: 22.6 cm
Bequest of Henry V. Putzel 9:1970

Although the diamond and boss is a common pattern among Shang and Zhou *gui*, the design is not especially well represented on bronzes in Western collections.[1] In America, this is one of about six early Western Zhou *gui* with handles.[2]

The mouth rim of the Museum's *gui* is beveled on the interior and exterior to produce a thick and rounded everted lip. On the interior, the mouth rim is further recessed to provide a reveal before the wall drops further down into the bowl of the vessel. The well-proportioned handles positioned below the flaring lip are aligned with the first of the three registers of decoration that adorn the body and foot. The low foot, a truncated cone with a narrow but sharply molded rim, is decorated with paired supine dragons which are contorted with bodies raised at diagonals, one above the other. The body of the *gui* bears a decoration of diamonds and bosses: three rows of round bosses in relief, the upper and lower rows formed by triangles with intaglio hooks. The bosses in the middle line are set in diamond lozenges with small hooks, bordered by a narrow field of rectangular *leiwen* and a deeply recessed line defining a larger diamond shape. The uppermost band is composed of a small animal mask flanked by three eyes with recessed pupils on a *leiwen* ground composed of interlocking S-curls.[3] The C-shaped handles are surmounted by animal heads with capped bottle horns. The headless bird bodies are delineated by raised hooks and small bosses, the tail and claws in intaglio on flat pendant hooks. Beneath each handle is a cicada motif, clearly a design element rather than an inscriptive mark.

The patina is a fine dark brown color bespeckled with light malachite and cuprite red and orange in a pitted surface. Chaplets are visible on the surface of the bronze.

The interior floor of the vessel is inscribed with four cast graphs in a single column. The characters may be read *Zuo Fu Yi Zi*, "Made for Father Yi by Zi."

The geometric pattern is derived from late Shang designs on round-bodied vessels, the diamonds and bosses of which are finely rendered, of smaller size, and more numerous in the use of pointed sharper bosses. In the Western Zhou, the diamonds and bosses are larger, fewer, and more smoothly rounded; the fewer the diamonds and bosses, the later the bronze. The use of this bold pattern and its strongly textural character continued in the early and middle Western Zhou.[4] By these criteria, the Museum's bronze is likely a product of the early Western Zhou dynasty.

In terms of its shape and decoration, there appears to be no exact parallel to the Saint Louis *gui*. While it is certainly a bowl, it lacks the rather squat and bulging body of the usual *gui* of the period; its walls are relatively vertical and steep with only a slight S-profile formed by the lip and body. The bronze is closest in the ornamentation of its first and second decorative registers to the same design bands of a *gui* in the Sackler collection.[5]

Notes
1. Rawson 1990, 378, cat. no. 41.

2. Hayashi Minao includes a number of diamond and boss *gui* with handles from the late Shang through Western Zhou in Western collections: Musée Cernuschi, Metropolitan Museum of Art, and the Musée Guimet (Hayashi 1984, 2:85, 109, 140). There are three in the Sackler collections (Rawson 1990, 370–79, 390–95, cat nos. 40–41, 45) and one at the Seattle Art Museum (*Haiwai* 1985, 12, pl. 12).

3. See Rawson 1990, 392–93 for a discussion of this register.

4. Rawson 1990, 374, 378; for chronologies of Western Zhou periods see ibid., 21.

5. Ibid., 390–95, cat. no. 45.

No. 33

Zhi

10th century BC
middle Western Zhou dynasty
Inscribed: *X*
Height: 12.5 cm
Width: 6.7 cm
Gift of J. Lionberger Davis
230:1954

The thin finished mouth rim of this elegant *zhi* tapers gently to a wide neck that broadens at the body to a generous waist. The high foot is accented by a short finished flange that is neatly undercut by a narrower plain foot with a flared finished base. The body and foot are elliptical in cross section. There are three areas of decor. At the neck and at the top of the elongated foot, the design is a single narrow *leiwen* band composed of a double row of interlocking scrolls. Below the flange on the foot is a simple decoration of two bowstrings.

Cast on the interior floor of the vessel is a figure resembling the reversed letter C and the capital letter L.[1]

The splotchy surface of malachite green is mixed with patches of smooth dark bronze; the interior of the mouth reveals relatively clean and smooth areas of light bronze color.

The plain and restrained character of the decoration on this bronze is comparable to an early Western Zhou *zhi* of slender proportions in the Sackler Collections.[2] This work with its wide, bulging body is of middle Western Zhou date. Although there seems to be no exact counterpart for the high double foot of this *zhi*, the penchant for high bases on several vessel types during the Western Zhou probably contributed to the elongated design and manufacture of this admittedly oddly shaped vessel.

Selected Bibliography
Kidder 1956, 34, pl. 2.

Notes
1. The inscription is thus far undecipherable and has prompted questions about the authenticity of this work. However, stylistically the bronze is comparable to works of the Western Zhou.

2. Rawson 1990, 630–31, cat. no. 101.

No. 34

Maguan

10th century BC
early Western Zhou dynasty
Height: 28 cm
Width: 41.8 cm
Museum Purchase
288:1949

This *maguan* is a large sculptural *taotie* with features rendered in multiple layers of low relief and openwork. The mask bears a prominent triple chevron crest and a broad pointed snout. The horns are large sharply pointed C-curls joined to smaller animal heads in profile; these heads have dimpled bosses for bulging eyes, capped bottle horns made of merged C-curls, pointed ears, and lower jaws. They bear a single clawed foot and a large broad plumelike appendage against which the relief of the heads stands out. The brows of the *taotie* are arching lunettes with broad borders filled with oblique fine striae of thin relief lines. The pointed ears of the mask are tightly curled with a cross of lines to form the aural canal. The eyes are comprised of a perforated boss surrounded by a striated ring and a larger and thicker plain ring. The mouth of the mask is upturned and flanked by large angled curling tusks.

The bronze is a mottled brown with gray-green malachite overlay. The snout and brows of the beast are encrusted with a rich malachite of darker hue, suggesting that these parts might have been treated differently from the rest of the mask and may have degraded in a different way as a result.

A smaller but comparable mask is in Stockholm.[1] Both masks exhibit chevrons at the forehead, brows textured with fine lines, everted protruding fangs, and concentrically formed eyes.[2]

Large *taotie* have been found on masks and equestrian gear, especially as chamfrons worn on the foreheads of horses, but none approaches the size of the Saint Louis piece. Early descriptions of excavations at the Western Zhou site of Xun Xian report "grotesque bronze masks, having some resemblance to human faces, were affixed to the south wall of the tomb, on either side of the approach."[3] There are three bronze non-structural straps running behind the eyes and across the top center of the rear of the mask which indicate that this work might have been attached to a background as decoration. This mask is said to have been found near Taiyuan, Shanxi.

Selected Bibliography
Rathbone 1950, 54; *Handbook* 1953, 254; Kidder 1956, 80, pl. 22.

Notes
1. Gyllensvärd and Pope 1966, 31, 128, cat. no. 25.

2. Some of these same features can be observed on the two masks in the Buckingham collection of The Art Institute of Chicago (Kelley and Ch'en 1946, 72–75, pls. 40–41).

3. Creel 1936, 248; Guo Baojun 1936 and 1964. For a discussion of bronze masks, see Meister 1938.

No. 35

Pan

mid 10th–early 9th century BC
middle Western Zhou dynasty
Inscribed: *Zuo Zhongfei zun Ge*
Height: 10.7 cm
Mouth diameter: 34.2 cm
Gift of J. Lionberger Davis Art Trust
126:1951

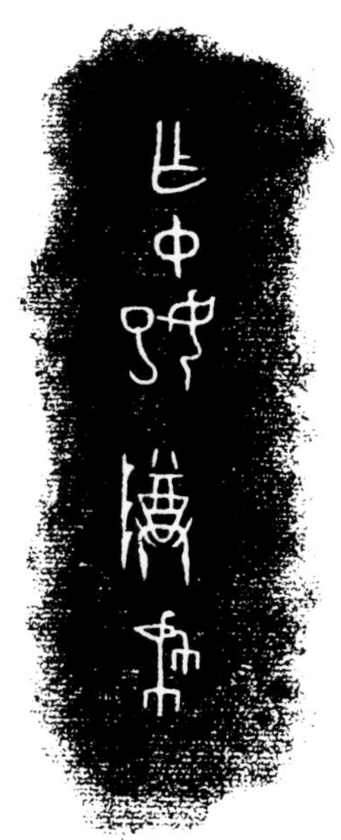

During the Western Zhou, the *pan* was used as a water container, a broad and shallow catch basin for water poured from another vessel. Commonly found in groups of bronzes excavated from Western Zhou sites, the *pan* is associated with two spouted vessel types, the *he* and the *yi*.[1] The *pan* with either the *he* or *yi* formed a set, sometimes with matched designs, for the purpose of washing hands.[2]

The molded and everted rim of this *pan* is beveled and slightly sloped towards the interior of the vessel. The body is short and straight-sided but rounded at the bottom where it constricts to a broad flared and molded foot. Two handles placed opposite one another protrude from the body just above the bottom of the vessel; the tops of the handles extend a short distance above the lip of the *pan* and butt against the rim. The exterior is decorated with a small antlered animal head in low relief, the snout of which extends below the band and into the narrow concave groove border. In four panels on either side of the raised *taotie* are pairs of addorsed diagonally supine dragons with high curving horns elaborated by recumbent C-shapes against a *leiwen* ground. The foot is decorated with eyes, volutes, and diagonals in intaglio.

The bronze bears a five-character inscription cast in a single column on the floor of the shallow bowl. The graphs may be read *Zuo Zhongfei zun Ge*, "Made for Zhongfei[3] [this] sacrificial [vessel], Ge." Ge, the pictograph of a halbard, is a clan name that had wide usage during the Shang and Zhou periods.[4]

A variety of S-shaped dragon designs and eye and diagonal motifs like those found on the Saint Louis *pan* can be found on bronzes of the middle Western Zhou.[5] The primary decor and the calligraphy of the Museum's *pan* are comparable to that of the Lu Fu Yu *pan*, an inscribed bronze of unknown provenance that was found in Xi'an, Shaanxi province in 1978. The Lu Fu Yu pan has been dated by inscription to the reign of the Zhou king Gong, around the second half of the 10th century BC.[6]

Selected Bibliography
Kidder 1956, 75–76, pl. 20; Barnard and Cheung 1978, 618, no. 790.

Notes
1. The incidence of *pan*, *he*, or *yi* found together in middle Western Zhou excavations include the archaeological sites in Shaanxi province at Dongjiacun, Qishan Xian (*Wenwu* 1976[5]:27– 44), Nanluo, Lintong (*Wenwu* 1982[1]:87–89), and Fufeng Xian (*Wenwu* 1976[6]:51–59).

2. Zhang Linsheng 1982.

3. Zhongfei literally means "middle consort." It may be read in context as "Lady Zhongfei."

4. Bagley 1987, 441–44, especially 442–43, cat. no. 79; Rawson 1990, 455, cat. no. 59.

5. Rawson 1990, 83–86, 584–85, cat. no. 89. Middle Western Zhou comparisons include the *pan* in the FitzWilliam Museum, Cambridge (Rawson 1990, 723, fig. 122.3).

6. Wang 1986, pl. 1.1.

No. 36

Hu

9th century BC
end of middle Western Zhou dynasty
Height: 44.8 cm
(with cover: 53.7 cm)
Foot diameter: 31.7 cm
Museum Purchase
281:1948

The broad, rhythmic, wavelike multiple bands encircling this *hu* appear in nascent form on late 10th-century bronzes excavated from Shandong province. There the undulating forms are used in high narrow single registers on the bodies or necks of their vessels.[1] The bronze designers' interest in a continuous kinetic design broken by vertical divisions (often called the wave pattern)[2] culminated in the development of the triple band arrangement and vessel shape of the Ji Fu *hu*, the exemplary model for the Saint Louis bronze.[3]

Slightly flared at the mouth rim, the *hu* constricts to a broad cylindrical neck that bears two lug handles and rings. The body of the vessel expands to a squat bulbous shape supported by a flaring and beveled foot. The cover is a low dome whose large flaring circular crown evolves into a beveled lip. The design on the vessel is three registers of spurred synclines and anticlines in low relief that resemble a continuous wave pattern encircling the neck and body. The wave in the top register is dished and marked by two parallel intaglio lines. Filling the trough of the waves are dished upright C-shapes above and pendant triangles below. The wave crests are ornamented with flat eye-shapes topped with three spikes; upright C-shapes fill the space below. The handles feature animal heads with spiral cone horns and curled nostrils in relief. The lower part of each lug bears intaglio curls and supports a hanging flat annular ring. The first and second registers are separated by a dished narrow band with a single centered intaglio line. The troughs of the waves in the second register are filled with similar dished upright C-shapes and pendant triangles; the crest of the waves are filled with upright triangles and a humpbacked animal with two heads in profile, with eyes of bosses and long proboscides. The interior of the body of the animal includes an eye-shape and two to three raised flared lines as supports. The waves of this second register are composed of two parallel dished and spurred synclines and anticlines. Separated from the second register by a plain dished narrow band, waves of the third register are comprised of two parallel dished and spurred synclines and anticlines plus an intervening grooved medial ridge that follows the dentations and flow of the waves. The troughs are filled with large, medium, and small upright C-shapes; the crests are filled with upright triangles and humpbacked, two-headed animals with upright elongated O-shapes in the bodies. Here and there, rudimentary and sometimes fragmentary *leiwen* fill the interstices. The foot is decorated with an eye and diagonal pattern. The cover, which bears wave patterns on its crown and lip, is cast on the top with a swirling long-plumed bird encircled by a band of cowry motifs.

The patina of the bronze is a uniform dark olive green. The surface is finely pitted and shows pale, light, and medium malachite green. There are patches of light gray-green corrosion of a leaden character on the vessel as well as traces of pale-beige earth. The color of the body and the lid are different. Indeed, the cover of the Museum's *hu* does not truly fit the mouth of the vessel, which suggests that it is not the original lid. The rings in the lug handles also may not be original.

In addition to being an important prototype in general shape and design, the Ji Fu *hu* also provides an eye and diagonal band at the foot which is neatly emulated in slightly altered form by the Museum's vessel.[4] The San Nian Xing *hu*, another excavated bronze from a late middle Western Zhou site in Shaanxi, similarly offers comparative motifs such as the swirling bird rondel and cowries cast on the lid and the curious circular motif surmounted by spikes.[5] The scant and disjointed but obvious use of *leiwen* on this and other wave-patterned *hu* appears to be related directly to the formative stages of the undulating design as represented by the late 10th-century vessels from Shandong, and later in the San Nian Xing *hu* where the wave patterns were set against a *leiwen* ground.[6] In terms of height and design, the closest comparison to the Saint Louis *hu* is a vessel in the British Museum which has scalloped overlapped cowries for rings, a band of cowry motifs at its foot, and no cover. In all other respects, the primary body decor of the British Museum *hu* is very similar to the present bronze, especially the upper register where the troughs of the waves are filled.[7]

Selected Bibliography
Kidder 1956, 82–84, pl. 24.

Notes
1. See the wave pattern laid against a ground of *leiwen* on the *zun* and *you* from Xiaoliuzhuang, Guicheng, Huang Xian, Shandong in Ji 1972, 5, pls. 7.1–2.

2. Jessica Rawson has traced the development of the wave pattern to the serial use of pendant triangular blades around bronzes, especially the *yu* vessel (Rawson 1990, 90–91, 455–59, cat. no. 59). Interest in a continuous rhythmic pattern can be traced at least to the Anyang phase of the Shang dynasty where potters employed geometric bands of linked chevrons, alternating plain and with *leiwen*, such as found in the famous white ceramic *lei* (39.42) in the Freer Gallery of Art (Lawton 1972, 173–74, cat. no. 85, pl. 85). During the middle Western Zhou, the older chevron design (Umehara 1951, nos. 21–22 and the Museum of Fine Arts, Boston, *hu* in Ackerman 1945, pl. 44) and the wave pattern both appear on slender covered *hu* (d'Argencé 1977, 94, cat. no. 39 back left).

3. The Ji Fu *hu* was found in 1960 at Qijiacun, Fufeng Xian, Shaanxi (Watson 1975, 81, cat. no. 94; *Xin Zhongguo* 1972, no. 51).

4. Watson 1975, 81, cat. no. 94; Shaanxi Sheng Bowuguan, et al. 1963, fig. 4.

5. See the San Nien Xing *hu* from the Wei Bo Xing group 1976 at Zhuangbai, Fufeng Xian, Shaanxi in *Shaanxi chutu* 1980, no. 31.

6. Ibid. For the Shandong vessels see Ji 1972, 5, pls. 7.1–2.

7. The British Museum and the Saint Louis *hu* are somewhat uncommon in that the troughs between the waves of the upper registers are filled. See the British Museum *hu* (1970.11–14.1; formerly Denis Cohen collection) standing 45.5 cm in height and dating from the 9th century BC in Rawson 1980, pl. 6. For an example of a wave pattern *hu* with an unfilled trough in the upper register see Lally 1992, cat. no. 22.

No. 37

Gui

first half of the 6th century BC
Eastern Zhou dynasty:
middle Spring and Autumn period
Height: 32.1 cm
Mouth diameter: 23.0 cm
Gift of J. Lionberger Davis
223:1950

Originally, this impressive bronze had an elaborate cover with a central ring of eight openwork petals and the same rhythmic pattern seen on the vessel and square base.[1] It is very similar to five other Eastern Zhou covered *gui* with square bases. This remarkable group of bronzes, which may have numbered as many as seven vessels from a single site, is now dispersed among collections in Asia and America.[2] Like the Saint Louis *gui*, each of the other bronzes is encompassed by a continuous wave pattern derived from earlier 10th- and 9th-century BC designs such as that found on the Museum's middle Western Zhou *hu* (No. 36; 281:1948). The archaistic use of this older, rather conservative pattern is extravagantly offset by the exuberant cover and the fabulously rampant dragons and beasts at the handles.

This *gui* is integrally cast with a high square pedestal. The flattened everted rim is sharply undercut to produce a distinctive recessed neck which is decorated with intaglio cowry decor. From the edge of the shoulder, the body curves out and then inward and down to the widely flaring foot. The vessel and stand are decorated with a system of synclines and anticlines of the wave pattern with addorsed C-shapes and triangular motifs alternately rising and pendant. The large handles are in the form of two dragons with swept-back horns and prominent fangs. The animals' heads are punctuated with large flat bosses and finely rendered intaglio lines and curls. Each is clutching to its chest a small animal with taloned forelegs and a large curling tail decorated with cowry designs.

The vessel's patina is a chalky pale green with large irregular patches of granular azurite that show smaller areas of tin oxide, cuprite, and malachite. Copper red also is revealed in the intaglio lines throughout the vessel, along with a general overlay of tin black accretions. The interior of the bowl is encrusted with azurite and the pseudomorphs of bird feathers. The bronze was cast in a four-part mold, the joins of which meet at the corners of the base. The dragons and beasts were cast separately and attached to the bowl by a small amount of molten bronze poured through holes in the shanks of the handles. The vessel has been heavily repaired, especially in the handles where radiographs reveal extensive restorations and modern structural inclusions. If the four others in the group are used as models, the dragons' crowns must have originally had a medial serpentine crest with hooks and spurs.

The wave pattern continued to be popular throughout the Western Zhou period. By the late 9th through the early 8th century the undulating design appeared in two forms: first, having dished and scooped ribbons with a sculpted character; and second, with broad flat ribbons created by intaglio lines. Openwork was *de rigueur*, occurring as crenelation on covers[3] or more subtly as high stems or feet.[4] In 6th-century examples such as the Museum's *gui*, the design elements, including the wave pattern, are rendered in thin lines of relief.

The Saint Louis vessel is directly related to the inscribed Qi Hou *yu*, a large and imposing bowl with two bands of wave patterns and four dragon handles and rings; it was discovered in 1957 in Luoyang, Henan province and weighs seventy-five kilograms.[5] The great bowl records the marriage of the second daughter of the duke of Qi to the Zhou king[6], and bears stiffly rendered wave patterns of a smaller scale that produce a more insistent patterning but also a more abstracted quality. The large bestial handles are bold but reductive and rather square in cross section. The Qi Hou *yu* is reckoned to have been created around 558 BC.[7] The softly modelled and sinuous forms of the dragon handles as well as the generous and more fluid treatment of the

waves on the Saint Louis *gui* and the other four identical vessels suggest a slightly earlier date than the Qi Hou *yu*.

Selected Bibliography
Loo 1941, cat. no. 77; Hoopes 1951, fig. 10; *Handbook* 1953, 255; Kidder 1956, 80–81, pl. 23.

Notes
1. The Museum's *gui* is shown as one of a pair of vessels, each with an openwork cover in Loo 1941, cat. no. 77.

2. The Saint Louis bronze is said to have been found with six other *gui* at a site in Linyi, Shandong province. Four of the other *gui* are known to the author; one of the vessels is said to be in Japan; the whereabouts of the seventh *gui* is unknown:

    a. Brundage collection (d'Argencé 1977, 104–5, cat. no. 44, front)

    b. Rockefeller collection (Mowry 1981, 50. [acc. no. 1979.103 a,b])

    c. Palace Museum collection, Beijing (*Zhongguo guqingtongqi* 1976, cat. no. 54, pl. 54)

    d. Cleveland Museum of Art collection (*Haiwai* 1985, 112, pl. 111)

    e. a fifth vessel in the "Nara Art Museum" [*sic*] recorded in d'Argencé 1977, 104 n. 7, cat. no. 44.

3. A pair of covered *fanghu* (I11M8:25–26) with elaborate handles and openwork wave patterns crowning the vessels was recently discovered in 1992 in a ducal tomb (I11M8), dating from around the reign of Xuan Wang (812–785 BC), near Beizhaocun, Qucunzhen, Quwo Xian, Shanxi province at Tianma, Qucun (*Wenwu* 1994[1]:4– 28).

4. See the openwork foot of a *dou* in *Wenhua* 1972, pl. 71, bottom.

5. So 1980, 264–65, 271, cat. no. 64, pl. 64; *Wenwu* 1960[4]:87; Zhang Jian 1977, 75, pls. 3.1–3.

6. Zhang Jian 1977.

7. Ibid.

Gui No. 37 (223:1950)

Yi

6th century BC
Eastern Zhou dynasty:
late Spring and Autumn period
Inscribed: *X Zi Shu Gou zi zuo yi yi
wannian yong zhi*
Height: 22.4 cm
Length: 38.5 cm
Gift of J. Lionberger Davis
213:1950

The four-legged *yi*, which had Shang prototypes as a tripod, first appears during the middle and late Western Zhou. In form and function, the *yi* is intimately related to the *gong* and *he* wine vessels of earlier times. Used in concert with the flat shallow *pan*, the *yi* was employed to pour water over the *pan* basin in the washing of hands.[1]

The spout of this *yi* is a deep U-shaped channel that rises above the rim of the body and handle. The tetrapod is supported by flat rectangular plaques well spaced under the body of the vessel. The handle is joined to the exterior rim of the body opposite the spout. The decorative horned bovine head has a stubby snout, its upper lip appearing pulled back from the lip of the container, which it bites; the protruding tail of the animal is curled and decorated with a small rope pattern. The handle is covered with an interlaced dragon design. The lower portion of the handle is tubular, an open hollow filled with core material. A wide band of repeated reversed S-shapes with large central curls and sharp spurs encircle the upper body and spout. A series of four smooth plain grooves flows around the container and meets under the mouth of the spout. The four legs are decorated with a fine pattern of triple vertical rows of intertwining dragons with tiny bosses for eyes, the curves of their bodies outlined in profile along the length of the leg. Surmounting each leg is a small *taotie* with tendril horns ending in sharp beaked heads.

The bronze has a dark reddish brown patina; the lower half of the vessel is covered with a fine corrosion of light olive green malachite with very small granular patches of azurite. The four legs are reinforced at the upper extremities by thickened bronze metal portions of square shape on the exterior and by long pointed shapes on the interior.

The floor of the vessel is inscribed with twelve finely cast and well-written characters in three columns of even-numbered graphs. The first character is illegible: the inscription *X Zi Shu Gou zi zuo yi yi wannian yong zhi,* may thus be read "**X** Zi Shu Gou made for himself this *yi* vessel, may it be used for ten thousand years."

The design of smooth deep grooves on the body of the vessel is derived from late Western Zhou bronzes when the technique and decoration were used not only on *yi* but more frequently on *gui*.[2] The minute interlocking designs on the upper body, legs, and particularly the handle are characteristic of 6th-century BC decorations.[3]

Selected Bibliography
"Additional Bronzes" 1951, 62; *Handbook* 1953, 255; Kidder 1956, 87–88, pl. 24; Barnard and Cheung 1978, 468, no. 450.

Notes
1. For a discussion on the use and development of the *yi* see Ma 1986, 202–3; Rawson 1990, 100–2; and especially Chen Fangmei 1989, 317.

2. See the grooved late Western Zhou *yi* from Qijiacun, Fufeng Xian, Shaanxi illustrated in Rawson 1990, 712–15, cat. no. 120, fig. 120.5; see Rawson 1990, 101, fig. 143b for drawings of the group of late Western Zhou bronzes, especially *yi* and *gui* from the finds at Shaochencun, Fufeng Xian, Shaanxi.

3. Compare the dragon designs on the Saint Louis *yi* handle with similar ornaments on a *ding* in Weber 1973, 109.

No. 39

He

4th century BC
Eastern Zhou dynasty:
middle Warring States period
Height: 23.4 cm
Bequest of Miss Leona J. Beckmann
16:1985

Three hipped and cabrioled legs support the squat rounded body of this covered *he*. The flat circular lid is connected by double ringed links to the overarching handle in elongated feline form. The short spout is modelled with the wings, head, and blunt beak of a bird. The body is divided into three decorated registers by two narrow concave grooves encircling the vessel. The decoration is rendered in very shallow intaglio so that the surface appears figured but smooth. The top of the cover is decorated with sinuous attenuated felines against a ground of large volutes and concentric *leiwen*. Around the flat side of the lid is an amorphous serpentine design in rather low relief which is divided into narrow rectangular panels. The three subsequent registers are filled with the forms of writhing dragons created with thin plain ribbons and filled with an array of figural devices.

The patina is a mottled light brown with reddish-blond highlights covered with a thin sugar-fine layer of light malachite green. Tin oxide and beige-colored earth appear in irregular patches throughout the vessel.

There are several features of the Saint Louis *he* that indicate a 4th-century BC or later date. The globular shape of the Museum's *he* is rounder than the squatter flattened shapes of typical vessels dating from the 5th and 4th centuries. Its spout is sited well within the first ornamental register and thus higher than spouts on earlier bronzes of the vessel type. In terms of the proportions and relationships of its separate elements to one another, the present bronze may be compared to the *he* of 5th–4th century date in the Buckingham Collection of The Art Institute of Chicago.[1]

Notes
1. So 1995, 411, fig. 84.3.

No. 40

Hu

4th–3rd century BC?
Eastern Zhou dynasty:
late Warring States period
Height: 29.2 cm
Mouth diameter: 12.6 cm
Museum Purchase 47:1918

This small but very fine object was among the first Chinese bronzes acquired by the Museum.[1] The vessel's minutely textured surface is a further abstraction of the intertwined dragon design. The small granular character of the decoration provides strong contrast to the plainer areas of the vessel, particularly the bands dividing one decorative register from the other. While the plethora of small raised curls and overlapping elements appears complex, if not chaotic, the decoration is given order and the shape of the vessel strengthened by the beautiful proportions of the ornamental registers.

The pair of lug handles that capture the vessel's two flat rings are capped by animal heads from whose mouths appear to issue the curving lower portion of the lug. These two curly-snouted animals with bulging eyes bear two horns created by parallel rows of cowry shell figures draped back and over the head, which is peppered with small raised dots. The vessel has a low foot with diagonal intaglio striations to simulate a twisted rope ring cradling a rounded base. The slightly flaring mouth of the *hu* tapers to a broad neck and then down to a restrained but generously swelling body. There are five registers of ornamentation. The plain neck and lower portion of the vessel below the belly are decorated with rising and descending lappets filled with interlocking C-curls in low relief and a horizontal element with two tiny bosses, a stamped pattern of devolved, intertwined dragons. The second, third, and fourth registers are broad bands of the same tightly interlocked low-relief design. The registers are divided by three narrow concave bands centered on a narrower raised cord of pointed devices; these elements are of equal length and alternate in dense and less dense intaglio designs.

The bronze has a lustrous patina of rich dark brown overlaid with a spidery patchwork of dark chocolate color. Dark green malachite occurs occasionally as surface corrosion; minute areas are filled with pale chalky green.

Selected Bibliography
Weber 1973, 321–25, cat. no. 58.

Notes
1. According to the dealer's invoice from the New York firm Dee and Fukushima, this bronze was formerly in the collection of Siegfried Bing (a.k.a Samuel Bing; 1838–1905), Paris. The eminent scholar Chen Mengjia first noted its "coarse" character in an examination of the piece on 14 May 1946, and wondered if the piece could be a copy. On 12–15 October 1993, W.T. Chase, Head Conservator of the Freer Gallery of Art/Sackler Gallery, remarked on its "rough" quality and suggested that it is "probably a lost-wax casting off of a genuine piece." On the basis of its design Weber suggested a "First Quarter of the Fifth Century" date (Weber 1973, 325, cat. no. 58).

No. 41

Ornament

3rd–2nd century BC
Qin–early Western Han dynasty
Height: 23.9 cm
Mouth diameter: 6.3 cm
Foot diameter: 7.3 cm
Museum Purchase
926:1920

This extraordinary inlaid tube is one of a group of five extant cylinders with silver designs. The bronzes vary in length and have differing fitted collars or flared ends, but all are identical in their designs and very nearly so in their dimensions.[1] In the past, these decorated metal cylinders have been called weapon butts, decorative sheaths for standards, and pole ornaments. They are now known to belong "to a set of chariot fittings used to join the handle of a canopy to a carriage."[2]

This bronze ornament is a thin-walled open cylindrical tube with a concave flaring rim at one end. Encircling the middle of the tube is a rounded ring dividing the main design into nearly equal halves. The design is vertically contiguous and symmetrical in arrangement, but is horizontally divided into six primary registers, excluding the narrow ornamental bands below the rim. The decoration is a complex design of silver and missing inlay: dots, circlets, starlets, pointed drops, volutes, dotted rosettes, bars of parallel lines, and recumbent curling S-shapes. The design is integrated by the orderly vertical symmetry as well as diagonal lines extending from one register to another.

The Saint Louis tube and the other four extant inlaid cylinders were probably part of a larger set which fitted together[3] as the ornamental sheathing of a wooden staff that supported the chariot's covering. The intricate and complex patterns of the inlaid fittings have been favorably compared to the designs on lacquers, textiles, and other bronze forms of the Western Han dynasty, especially the spectacular tomb of the wife of the Marquis of Dai (ca.150 BC) at Mawangdui, Changsha, Hunan province.[4]

Selected Bibliography
Kidder 1956, 92–93, pl. 29; Loehr 1968, 172–73, cat. no. 79.

Notes
1. The four other sections, single and double, may be found in the following references: two separate sections reproduced in Yetts 1929, 63, cat. nos. A99–100; one section in Rawson 1980, pl. 12; and two joined sections in Deydier 1991, 48–50, cat. no. 12. The Deydier 1991 piece is the longest and most beautifully preserved of all the inlaid tubes. Compare the similar but different two tubes illustrated in Hansford 1957, 69, cat. nc. A52, pl. 26.

2. Wu Hung 1984, 38 n. 7

3. The single section of inlaid tube in the British Museum shows an interior raised and bevelled collar meant to fit into another length of tubing (Rawson 1980).

4. Ibid., 219; Rawson and Bunker 1990, 184, cat. no. 94; Wu Hong 1984, 48.

REFERENCES

Ackerman 1945	Ackerman, Phyllis. 1945. *Ritual Bronzes of Ancient China*. New York: Dryden Press.
"Additional Bronzes" 1951	"Additional Bronzes from the J. Lionberger Davis Gift." 1951. *Bulletin of the City Art Museum of St. Louis* 36(4):58–63.
Allen 1950	Allen, Maude Rex. 1950. "Early Chinese Lamps." *Oriental Art* 2(4):133–40.
d'Argencé 1977	d'Argencé, René-Yvon Lefebvre. 1977. *Bronze Vessels of Ancient China in the Avery Brundage Collection*. San Francisco: Asian Art Museum of San Francisco.
Bagley 1980	Bagley, Robert W. 1980. "The Rise of the Western Zhou Dynasty." In *The Great Bronze Age of China*, ed. Wen Fong, 191–238. New York: The Metropolitan Museum of Art.
Bagley 1987	———. 1987. *Shang Ritual Bronzes in the Arthur M. Sackler Collections*. Ancient Chinese Bronzes in the Arthur M. Sackler Collections, vol. 1. Cambridge: Harvard University Press.
Barnard and Cheung 1978	Barnard, Noel, and Cheung Kwong-yue. 1978. *Zhong Ri Ou Mei Aoniu suojian suota suomo jinwen huibian/Rubbings and Handcopies of Bronze Inscriptions in Chinese, Japanese, European, American, and Australian Collections*. 10 vols. Taipei: Yee Wen Publishing Company.
"A Bronze Sacrificial Vessel" 1916	"A Bronze Sacrificial Vessel of the Chou Dynasty in China." 1916. *Bulletin of the City Art Museum of St. Louis* 2(3):9.
Buchanan and Ronan 1980	Buchanan, FitzGerald, and Colin A. Ronan. 1980. *China: The Land and the People*. New York: Crown Publishers.
Cao 1980	Cao Dingyun. 1980. "Ya qi kao." *Wenwu jikan* (2):143–50.
Chase 1981	Chase, W. Thomas. 1981. "Bronze Casting in China: A Short Technical History." In *The Great Bronze Age of China: A Symposium*, ed. George Kuwayama, 100–23. Los Angeles: Los Angeles County Museum of Art.
Cheng, et al. 1989	Cheng Changxin, Wang Yongchang, and Cheng Ruixiu. 1989. "Tongqi bianwei qianshuo (zhong)." *Wenwu* (11):33–44.
Chen Fangmei 1989	Chen Fangmei. 1989. *Shang Zhou qingtong jiuqi tejan tulu/Catalogue of the Special Exhibition: Shang and Chou Dynasty Bronze Wine Vessels*. Taipei: National Palace Museum.
Chen and Zhang 1984	Chen Gongrou and Zhang Changshou. 1984. "Yin Zhou qingtongqi shang niaowen de duandai yanjiu." *Kaogu xuebao* (3):265–85.
Chen Jieqi 1918	Chen Jieqi. 1918. *Fuzhai jijin lu*. Ed. Teng Shi. N.p.: Fengleilou.
Chen Mengjia 1946	Chen Mengjia. 1946. "Style of Chinese Bronzes." *Archives of the Chinese Art Society of America* 1:26–52.
Chen Mengjia 1956	———. 1956. *Yinxu buci zongshu*. Beijing: Kexue Chubanshe.
Chen Mengjia 1977	———, ed. 1977. *In Shū seidōki bunrui zuroku*. 2 vols. Tokyo: Kyūko Shorin.
Chen Zhengxiang 1975	Chen Zhengxiang. 1975. *Zhongguo gujin diming da cidian*. Taipei Shangwu Yinshuguan.
Chêng 1956	Chêng Tê-k'un. 1956 "The Ya Lü-Kuei." *Oriental Art* 2(1):9–16.
Chêng 1957	———. 1957. "Review of Early Chinese Bronzes in the City Art Museum of St. Louis." *Oriental Art* 3(1):34–35.
Chêng 1965	———. 1965. "The T'u-lu Color-Container of the Shang-Chou Period." *Bulletin of the Museum of Far Eastern Antiquities, Stockholm* (37):239–50.
"Chinese Art" 1951	"Chinese Art Recently Acquired by American Museums." 1951. *Archives of the Chinese Art Society of America* 5:70–77.
"Chinese Art" 1952	"Chinese Art Recently Acquired by American Museums." 1952. *Archives of the Chinese Art Society of America* 6:64–72.
Chu 1983	Chu Ko. 1983. "Ren shou zhi jian." *Gugong wenwu yuekan* 1(3):34– 51.

| Consten 1958 | Consten, Eleanor von Erdberg. 1958. *Das alte China*. Stuttgart: Gustav Kilpper Verlag. |

Consten 1958 — Consten, Eleanor von Erdberg. 1958. *Das alte China*. Stuttgart: Gustav Kilpper Verlag.

Creel 1936 — Creel, Herrlee Glessner. 1936. *The Birth of China: A Survey of the Formative Period of Chinese Civilization*. London: Jonathan Cape.

Davidson 1937 — Davidson, Leroy. 1937. "Toward a Grouping of Early Chinese Bronzes." *Parnassus* 4(4):29–34, 51.

Davis 1956 — Davis, Frank. 1956. "A Page for Collectors: A Magnificent Gift to Oxford University." *The Illustrated London News*, 22 September:472–74.

Découverte 1954 — *La Découverte de l'Asie: Hommage à René Grousset*. 1954. Paris: Musée Cernuschi.

Deydier 1980a — Deydier, Christian. 1980. *Archaic Chinese Bronzes from Shang and Zhou Dynasties*. London: Oriental Bronzes Ltd.

Deydier 1980b — ———. 1980. *Chinese Bronzes*. Trans. Janet Seligman. New York: Rizzoli International.

Deydier 1991 — ———. 1991. *The Art of the Warring States and Han Periods*. London: Oriental Bronzes Ltd.

Deydier 1995 — ———. 1995. *Les Bronzes archaïques chinois, Xia et Shang*. Paris: Les Éditions d'Art et d'Histoire.

Du 1994 — Du Jinpeng. 1994. "Shang Zhou tong jue yanjiu." *Kaogu xuebao* (3):263–98.

Duanfang 1908 — Duanfang. 1908. *Taozhai jijin lu*. Shanghai: Yuzheng Shuju.

Ecke 1939 — Ecke, Gustav. 1939. *Frühe chinesische Bronzen aus der Sammlung Oskar Trautmann*. Beijing: Furen University.

An Exhibition 1940 — *An Exhibition of Ancient Chinese Ritual Bronzes*. 1940. Detroit: Detroit Institute of Arts.

"The Exhibition" 1934 — "The Exhibition of Early Chinese Bronzes." 1934. *Bulletin of the Museum of Far Eastern Antiquities, Stockholm* (6):81–136.

Fang, et al. 1993 — Fang Shuxin, et al., eds. 1993. *Jiagu jinwen zidian*. Chengdu: Ba Shu Shushe.

Fine Chinese Ceramics 1969 — *Fine Chinese Ceramics, Lacquers, and Bronzes*. 1969. New York: Christie's of London.

Fong 1980 — Fong, Wen, ed. 1980. *The Great Bronze Age of China: An Exhibition from the People's Republic of China*. New York: The Metropolitan Museum of Art.

Gao Ming 1992 — Gao Ming. 1992. "Xi Zhou jinwen 'X' zi ziliao zhengli he yanjiu." In *Kaoguxue yanjiu*. Bejing: Beijing Daxue Kaoguxue Congshu.

Gao Quxun 1962 — Gao Quxun, ed. 1962. *Houjiazhuang (Henan Anyang Houjiazhuang Yin dai mudi) dier ben, 1001 hao damu* (vol. 2, pts. 1–2). Archaeologia Sinica, no. 3. Nangang: Academia Sinica.

Gao Quxun 1967 — ———, ed. 1967. *Houjiazhuang (Henan Anyang Houjiazhuang Yin dai mudi) disi ben, 1003 hao damu* (vol. 4). Archaeologia Sinica, no. 3. Nangang: Academia Sinica.

Ge 1972 — Ge Jin. 1972. "Jingyang Gaojinpao zao Zhou mucang fajueji." *Wenwu* (7):3–18.

Goepper 1972 — Goepper, Roger, ed. 1972. *Form und Farbe: Chinesische Bronzen und Frühkeramik Sammlung H.W. Siegel*. Cologne: Museum für Ostasiatische Kunst der Stadt Köln.

Gugong 1958 — *Gugong tongqi tulu*. 1958. 2 vols. Taipei: Taiwan Shudian.

Guo Baojun 1936 — Guo Baojun. 1936. "Xun Xian Xincun gucanmu zhi qinglu." *Tianyie kaogu paogao* (1):167–200.

Guo Baojun 1964 — ———. 1964. *Xun Xian Xincun*. Beijing: Kexue Chubanshe.

Guo Moruo 1952 — Guo Moruo. 1952. *Jinwen congkao*. Rev. ed. Beijing: Renmin Chubanshe.

Gyllensvärd and Pope 1966 — Gyllensvärd, Bo, and John A. Pope. 1966. *Chinese Art from the Collection of H.M. King Gustaf VI Adolf of Sweden*. New York: The Asia Society.

Haiwai 1985 — *Haiwai yizhen: Tongqi/Chinese Art in Overseas Collections: Bronze*. 1985. Taipei: National Palace Museum.

Haiwai 1988 — *Haiwai yizhen: Tongqi (xu)/Chinese Art in Overseas Collections: Bronze (II)*. 1988. Taipei: National Palace Museum.

Handbook 1953   *The City Art Museum of St. Louis Handbook of the Collections.* 1953. St. Louis: City Art Museum of St. Louis.

Handbook 1975   *The St. Louis Art Museum Handbook of the Collections.* 1975. St. Louis: The St. Louis Art Museum.

Hansford 1957   Hansford, S. Howard. 1957. *The Seligman Collection of Oriental Art.* Vol. 1. London: Lund Humphries.

Hayashi 1984   Hayashi Minao. 1984. *In Shū jidai seidōki no kenkyū: In Shū seidōki sōran ichi.* 2 vols. Tokyo: Yoshikawa Kōbunkan.

Henan chutu 1981   *Henan chutu Shang Zhou qingtongqi.* 1981. Vol. 1. Beijing: Wenwu Chubanshe.

Hentze 1951   Hentze, Carl. 1951. *Bronzegerät, Kultbauten, Religion im ältesten China der Shang-zeit.* 2 vols. Antwerp: De Sikkel.

van Heusden 1952   van Heusden, Willem. 1952. *Ancient Chinese Bronzes of the Shang and Chou Dynasties: An Illustrated Catalogue of the van Heusden Collection with a Historical Introduction.* Tokyo: P.p.

Higuchi 1973   Higuchi Takayasu, ed. 1973. *Chūgoku bijutsu.* Tokyo: Kōdansha.

Higuchi and Enjōji 1984   Higuchi Takayasu and Enjōji Jirō. 1984. *Chūgoku seidōki hyakusen.* Tokyo: Nihon Keisai Shinbunsha.

Hoopes 1951   Hoopes, Thomas T. 1951. "Gift of Chinese Ritual Bronzes." *Bulletin of the City Art Museum of St. Louis* 36(1): cover, 2–7.

Huang 1937   Huang Jun. 1937. *Yezhong pianyu er ji.* Beijing: Zuguzhai.

Huang 1942   ———. 1942. *Yezhong pianyu san ji.* Beijing: Zuguzhai.

Huber 1981a   Huber, Louisa G. 1981. "Ancient Chinese Bronzes." *Arts of Asia* 11(2):74–87.

Huber 1981b   Huber, Louisa G. Fitzgerald. 1981. "Some Anyang Royal Bronzes: Remarks on Shang Bronze Decor." In *The Great Bronze Age of China: A Symposium*, ed. George Kuwayama, 16–43. Los Angeles: Los Angeles County Museum of Art.

Hummel 1943   Hummel, Arthur W., ed. 1943. *Eminent Chinese of the Ch'ing Period.* Washington, D.C.: Library of Congress.

Hunan Sheng Bowuguan 1983   *Hunan Sheng Bowuguan.* 1983. Beijing: Wenwu Chubanshe.

Important Chinese Works of Art 1983   *Important Chinese Works of Art: The Collection of Mr. and Mrs. Richard Bull.* 1983. Sale 5122. New York: Sotheby Parke Bernet.

Important Early Chinese Dynastic Bronzes 1948   *Important Early Chinese Dynastic Bronzes Collected by the Late James Monroe McAdams.* 1948. Sale 1000. New York: Parke-Bernet Galleries.

Ji 1972   Ji Wenshou. 1972. "Kaishu jinnianlai Shandong chutudi Shang Zhou qingtongqi." *Wenwu* (5):5–8.

Kane 1970   Kane, Virginia. 1970. "Chinese Bronze Vessels of the Shang and Western Zhou Periods." Ph.d. diss., Harvard University.

Kane 1973   ———. 1973. "The Chronological Significance of the Inscribed Ancestor Dedication in the Periodization of Shang Dynasty Bronze Vessels." *Artibus Asiae* 35(4):335–70.

Kaogu.   Beijing: Kexue Chubanshe, 1959–present.

Kaogu xuebao.   Beijing: Kexue Chubanshe, 1951–present.

Karlgren 1936   Karlgren, Bernhard. 1936. "Yin and Chou in Chinese Bronzes." *Bulletin of the Museum of Far Eastern Antiquities, Stockholm* (8):9–154.

Karlgren 1937   ———. 1937. "New Studies on Chinese Bronzes." *Bulletin of the Museum of Far Eastern Antiquities, Stockholm* (9):9–118.

Karlgren 1952   ———. 1952. *A Catalogue of the Chinese Bronzes in the Alfred M. Pillsbury Collection.* Minneapolis: Minneapolis Institute of Arts.

Karlgren 1962   ———. 1962. "Some Characteristics of the Yin Art." *Bulletin of the Museum of Far Eastern Antiquities, Stockholm* (34):1–28.

Keightley 1978 — Keightley, David N. 1978. *Sources of Shang History: The Oracle-Bone Inscriptions of Bronze Age China*. Berkeley, Los Angeles, and London: University of California Press.

Kelley and Ch'en 1946 — Kelley, Charles Fabens, and Ch'en Meng-chia (Chen Mengjia). 1946. *Chinese Bronzes from the Buckingham Collection*. Chicago: The Art Institute of Chicago.

Kidder 1956 — Kidder, J. Edward, Jr. 1956 *Early Chinese Bronzes in the City Art Museum of St. Louis*. St. Louis: City Art Museum of St. Louis.

Kim 1940 — Kim, Chewon. 1940. "A Chinese Bronze Lamp." *Parnassus* 12(2):23– 24.

Koyama, et al. 1975 — Koyama Fujio, Sugimura Yūzō, Sekino Takeshi, and Miyagawa Torao, eds. 1975. *Kokyū Hakubutsuin*. 2 vols. Tokyo: Kōdansha.

Lally 1992 — Lally, James J. 1992. *Chinese Archaic Bronzes: Sculpture and Works of Art*. New York: J.J. Lally & Company.

Lally 1994 — ———. 1994. *Archaic Chinese Bronzes: Jades and Works of Art*. New York: J.J. Lally & Company.

Lawton 1972 — Lawton, Thomas. 1972. *The Freer Gallery of Art, Part I: China*. Tokyo: Kōdansha.

Li 1948 — Li Ji. 1948. "Xiaotun chutu zhi qingtongqi." *Zhongguo kaogu xuebao* 3:1–100.

Li 1956 — ———. 1956. *Xiaotun (Henan Anyang Yinxu yizhi zhi yi) disan ben, Yinxu qiwu: jia bian, taoqi: shang ji* (vol. 3, fasc. 1, pt. 1). Archaeologia Sinica, no. 2. Nangang: Academia Sinica

Li and Wan 1968 — Li Ji and Wan Jiabao. 1968. *Yinxu chutu qingtong jia xing qi zhi yanjiu*. Nankang: Academia Sinica.

Li and Wan 1972 — ———. 1972. *Yinxu chutu wushisan jian qingtong rongqi zhi yanqiu*. Nangang: Academia Sinica.

Lienert 1979 — Lienert, Ursula. 1979. *Typology of the Ting in the Shang Dynasty: A Tentative Chronology of the Yin-hsü Period*. Publikationen der Abteilung Asien Kunsthistorisches Institut der Universität Köln, vols. 3.1–2. Weisbaden: Franz Steiner Verlag.

Liu 1902 — Liu Xinyuan. 1902. *Qigushi jijin wenshu*.

Loehr 1968 — Loehr, Max. 1968. *Ritual Vessels of Bronze Age China*. New York: The Asia Society.

Loo 1941 — Loo, C.T. 1941. *Exhibition of Chinese Arts*. New York, C.T. Loo & Company.

Luo 1937 — Luo Zhenyu. 1937. *Sandai jijin wencun*. 20 vols. Beijing.

Ma 1986 — Ma Chengyuan. 1986. *Ancient Chinese Bronzes*. Ed. Hsio-yen Shih. Hong Kong: Oxford University Press.

Meister 1938 — Meister, P.W. 1938. "Chinesische Bronzemasken." *Ostasiatische Zeitschrift* 14(1):5–11.

Mizuno 1958 — Mizuno Seiichi, ed. 1958. *Sekai kōkogaku taikei 6: In Shū jidai*. Tokyo: Heibonsha.

Moffitt and Voris 1988 — Moffitt, John F., and Peter Voris. 1988. "The 'Feng' Peacock: An Emblematic Device on Early Zhou Vessels." *Arts of Asia* 18(4) 104–11.

Mortimer 1985 — Mortimer, Kristen A. 1985. *Harvard University Art Museums: A Guide to the Collections*. Cambridge: Harvard University Art Museums.

Mowry 1981 — Mowry, Robert. 1981. *Handbook of the Mr. and Mrs. John D. Rockefeller 3rd Collection*. New York: The Asia Society.

Murray 1985 — Murray, Julia K. 1985. "Object of the Month: Early Western Zhou Bronze Fang Yi." *Orientations* 16(10):42–44.

"A New Installation" 1917 — "A New Installation." 1917. *Bulletin of the City Art Museum of St. Louis* 3(3):11–12.

Ostasiatische Kunst 1992 — *Ostasiatische Kunst*. 1992. Cologne: Kunsthaus Lempertz.

Pope, et al. 1967 — Pope, John Alexander, et al. 1967. *The Freer Chinese Bronzes*. 2 vols. Washington, D.C.: Freer Gallery of Art.

Rathbone 1950 — Rathbone, Perry. 1950. "An Early Chinese Bronze Mask." *Bulletin of the City Art Museum of St. Louis* 35(4):54–55.

Rawson 1980   Rawson, Jessica. 1980. *Ancient China: Art and Archaeology*. London: British Museum.

Rawson 1984   ———. 1984. "Eccentric Bronzes of the Early Western Zhou." *Transactions of the Oriental Ceramic Society, 1982–1983*, 10–31. London: The Oriental Ceramic Society, 1984.

Rawson 1987   ———. 1987. *Chinese Bronzes: Art and Ritual*. London: British Museum.

Rawson 1990   ———. 1990. *Western Zhou Ritual Bronzes from the Arthur M. Sackler Collections*. Ancient Chinese Bronzes in the Arthur M. Sackler Collections, vols. 2A–B. Cambridge: Harvard University Press.

Rawson 1993   ———. 1993. "Late Shang Bronze Design: Meaning and Purpose." *The Problem of Meaning in Early Chinese Bronzes*. London: School of Oriental and African Studies.

Rawson and Bunker 1990   Rawson, Jessica, and Emma C. Bunker. 1990. *Ancient Chinese and Ordos Bronzes*. Hong Kong: The Oriental Ceramic Society of Hong Kong.

Rong 1941   Rong Geng. 1941. *Shang Zhou yiqi tongkao*. 2 vols, Beijing: Harvard-Yenching Institute.

Ruan 1797   Ruan Yuan. 1797. *Shanzuo jinshi zhi*. Yizheng: Xiaolanghuanxianguan.

Salles 1934   Salles, Georges. 1934. *Bronzes Chinois des Dynasties Tcheou, Ts'in et Han*. Paris: Musée de l'Orangerie.

Shaanxi Sheng Bowuguan, et al. 1963   Shaanxi Sheng Bowuguan, et al. 1963. *Fufeng Qijiacun qingtongqi qun*. Beijing: Wenwu Chubanshe.

Shaanxi chutu 1979   *Shaanxi chutu Shang Zhou qingtongqi*. 1979. Vol. 1. Beijing: Wenwu Chubanshe.

Shaanxi chutu 1980   *Shaanxi chutu Shang Zhou qingtongqi*. 1980. Vol. 2. Beijing: Wenwu Chubanshe.

Shang 1935   Shang Chengzuo, ed. 1935. *Shierh jia jichin tulu*. Shanghai: Commercial Press.

Shanghai Bowuguan 1985   *Shanghai Bowuguan*. 1985. Beijing: Wenwu Chubanshe.

So 1980   So, Jenny. 1980. "New Departures in Eastern Zhou Bronze Designs: The Spring and Autumn Period." In *The Great Bronze Age of China*, ed. Wen Fong, 249–301. New York: The Metropolitan Museum of Art.

So 1995   ———. 1995. *Eastern Zhou Ritual Bronzes from the Arthur M. Sackler Collections*. Ancient Chinese Bronzes in the Arthur M. Sackler Collections, vol. 3. New York: Abrams.

Somogyi 1986   Somogyi, Lisa L. 1986. "An Exploration of the Concept of 'Archaism' within a Study of Three Ritual Vessels of the Late Shang Period in The St. Louis Art Museum." Master's thesis, Washington University.

Thorp 1988   Thorp, Robert L. 1988. "The Archaeology of Style at Anyang: Tomb 5 in Context." *Archives of Asian Art* 41:45–69.

Umehara 1947   Umehara Sueji. 1947. *Kankarō kikkinzu*. Kyoto: P.p.

Umehara 1951   ———. 1951. *Hakutsuru kikkin senshū*. Kobe: Hakutsuru Bijutsukan.

Umehara 1959   ———. 1959. "Sensei-shō Hōkei-ken shutsudo no dai ni no henkin." *Tōhōgaku kiyō* 1:1–15.

Umehara 1959–64   ———. 1959–64. *Nihon shūcho Shina kodō seika*. 6 vols. Osaka: Yamanaka Shōkai.

Umehara 1964   ———. 1964. *Inkyo/Yin Hsu: Ancient Capital of the Shang Dynasty at An-yang*. Tokyo: Asahi Shinbunsha.

Wang 1986   Wang Shangxing. 1986. "Lu Fu Yu pan ming kaoshi ji qi xiangkuan wenti." *Wenwu* (4):1–7.

Watson 1962   Watson, William. 1962. *Ancient Chinese Bronzes*. London: Faber and Faber.

Watson 1965   ———. 1965. *Foundations of Chinese Art*. London: Thames and Hudson.

Watson 1975   ———. 1975. *The Chinese Exhibition: The Exhibition of Archaeological Finds of the People's Republic of China*. Kansas City: Nelson Gallery-Atkins Museum.

Watt 1990   Watt, James C.Y. 1990. "The Arts of Ancient China." *The Metropolitan Museum of Art Bulletin* 48(1):1–72.

Weber 1973   Weber, George W., Jr. 1973. *The Ornaments of Late Chou Bronzes: A Method of Analysis*. New Brunswick: Rutgers University Press.

Wendel 1986   Wendel, Katherine. 1986. "An Analysis of Archaism in Fang-ting Vessels of the Late Shang." Master's thesis, Washington University.

Wenhua 1972   *Wenhua da geming qijian chutu wenwu, diyi ji*. 1972. Beijing: Wenwu Chubanshe.

Wenwu.   Beijing: Wenwu Chubanshe, 1950–present.

Wu Hung 1984   Wu Hung. 1984. "A Sanpan Shan Chariot Ornament and the Xiangrui Design in Western Han Art." *Archives of Asian Art* 37:37–59.

Wu Hung 1995   ———. 1995. *Monumentality in Early Chinese Art and Architecture*. Stanford: Stanford University Press.

Wu Shifen 1895   Wu Shifen. 1895. *Jungu lu jinwen*. Kaifeng.

Xiaotun 1985   *Xiaotun*. Beijing: Institute of Archaeology, 1985.

Xin Zhongguo 1972   *Xin Zhongguo chutu wenwu*. 1972. Beijing: Waiwen Chubanshe.

Xu Jinxiung 1968   Xu Jinxiung. 1968. *Yin buci zhong wuzhong jisi de yanjiu*. Taipei: Taiwan University.

Xu Shen 1975   Xu Shen. 1975. *Shuowen jiezi kulin ji buyi*. 16 vols. Ed. Ding Fubao. Taipei: Taipei Shangwu Yinshu Guan.

Xu Shen n.d.   ———. n.d. *Shuowen jiezi*. In *Siku shanben congshu chubian*, ed. Ji Yun. Beijing: Zhongyang Tushuguan.

Yan 1975   Yan Wan. 1975. "Beijing Liaoning chutu tongqi Zhou chu di Yan." *Kaogu* (5):274–75.

Yang 1965–66   Yang Xiong. 1965–66. *Tai Xuan Jing*. 10 vols. Sibu beiyao, no. 357. Taipei: Zhonghua Shuzhu.

Yetts 1929   Yetts, William Perceval. 1929. *The George Eumorfopoulos Collection: Catalogue of the Chinese and Corean Bronzes, Sculpture, Jade and Jewelry, and Miscellaneous Objects, Vol. 1: Bronzes: Ritual and Other Vessels, Weapons, Etc.* London: Ernest Benn Ltd.

Yetts 1939   ———. 1939. *The Cull Chinese Bronzes*. London: University of London, Courtauld Institute of Art.

Yinxu 1985   *Yinxu qingtongqi*. 1985. Beijing: Wenwu Chubanshe.

Zhang Jiyun 1973   Zhang Jiyun, ed. 1973. *Zhongwen da cidian*. 10 vols. Taipei: China Academy.

Zhang Jian 1977   Zhang Jian. 1977. "Qi Hou jian mingwen di xin faxian." *Wenwu* (3):75.

Zhang Linsheng 1982   Zhang Linsheng. 1982. "Shuo he yu yi: Qingtong yiqi zhongde shuiqi." *Gugong jikan* 17(1):25–39.

Zhang Shui 1972   Zhang Shui. 1972. "Jishan Jiajiacun chutudi Xi Zhou tongqi." *Wenwu* (6):25–27.

Zhang and Liu 1981   Zhang Yachu and Liu Yu. 1981. "Cong Shang Zhou bagua shuzi fuhao tan shifa di jige wenti." *Wenwu* (2):154, 155–63.

Zhang and Liu 1981–82   ———. 1981–82. "Some Observations About Milfoil Divination Based on Shang and Zhou *Bagua* Numerical Symbols." Trans. Edward L. Shaughnessy. *Early China* 7:46–55.

Zhang Zhenglang 1980   Zhang Zhenglang. 1980. "Shishi Zhouchu qingtongqi mingwen zhong de Yigua." *Kaogu xuebao* (4):403–15.

Zhang Zhenglang 1980–81   ———. 1980–81. "An Interpretation of the Divinatory Inscriptions on Early Zhou Bronzes." Trans. Horst W. Huber, Robin D.S. Yates, et al. *Early China* 6:80–96.

Zhongguo guqingtongqi 1976   *Zhongguo guqingtongqi xuan*. 1976. Beijing: Wenwu Chubanshe.

Zou 1980   Zou Heng. 1980. "Guanyu Xia Shang shiqi beifang diqu Zhuling qu wenhua chubu shentao." In *Xia Shang Zhou kaoguxue lunwenji*, 253–94. Beijing: Wenwu Chubanshe.

Zhou, et al. 1974–75   Zhou Fagao, Li Xiaoding, and Zhang Risheng, eds. 1974–75. *Jinwen gulin*. 16 vols. Hong Kong: The Chinese University of Hong Kong.

Zhou, et al. 1977   ———, eds. 1977. *Jinwen gulin fulu/Appendixes to an Etymological Dictionary of Ancient Chinese Bronze Inscriptions*. Hong Kong: The Chinese University of Hong Kong.

Technical Observations and Analyses of Chinese Bronzes

Suzanne Hargrove

The Chinese bronzes represented in this catalogue are a culmination of the considerable skills of ancient craftsmen, alterations wrought by time and burial, and subsequent changes resulting from treatment by individuals including archaeologists, dealers, restorers, and conservators. Technical studies of the bronzes elucidate aspects of the technology employed in manufacturing them, as well as an identification of all that has happened to them since they were made.

This is an exciting albeit challenging task, involving a separation of fact from fiction that is by no means definitive. Consideration must be given to the original, present, and actual condition of each bronze in describing their appearance as they exist today.[1] The original condition refers to the determination of the object's probable appearance when it was created and used, based on existing features identified in a detailed examination and knowledge gained from archaeological, conservation, and art historical records. The present condition describes the object's current state, including alteration products from burial, repairs, replacement of lost and damaged parts, artistic embellishment (e.g., designs, appendages), removed material (e.g., corrosion products), and applied products (e.g., coatings, fills, paint). Authentication of the work and/or its various parts would also be considered. The actual condition refers to how much of the bronze remains and its state of existence.

A number of tools have been utilized to accomplish this task, including a detailed visual examination; close inspection of surface features under stereomicroscope; solvent testing of materials applied to the surface; analytical sampling to identify materials using X-ray powder diffraction, lead isotope analysis, infrared spectroscopy, and cross section analysis; and nondestructive diagnostic procedures such as X-ray fluorescence, examination under long-wave ultraviolet light, X-radiography, and X-ray computed tomography.

The main components used to create Chinese bronzes consist of the mold, model (optional), core, and chaplets (Diagrams 1–2).

Chinese bronze vessels were created exclusively by pouring molten metal into stone or ceramic piece-molds.[2] A mold is a receptacle that defines the shape, decoration, and surface finish of the vessel, and therefore contains a negative impression of the vessel. The model (positive) is a solid three-dimensional representation of the bronze done prior to casting. Models are used to transfer design features to the mold. Whether models were a necessary part of bronze fabrication is debatable.[3]

Both mold and model were probably made of loess.[4] When a model was used, it might have been baked to harden it in preparation for taking mold impressions.[5] To form the mold, a finer grade of loess would be pressed against the model in a thin layer to obtain fine details, followed by a coarser grade applied as an outer support layer. This impression would be removed while in a leather hard state to allow for additional touch-up or decoration. The mold would be divided into sections to facilitate its removal from the model (and eventually the cast bronze), and to allow direct

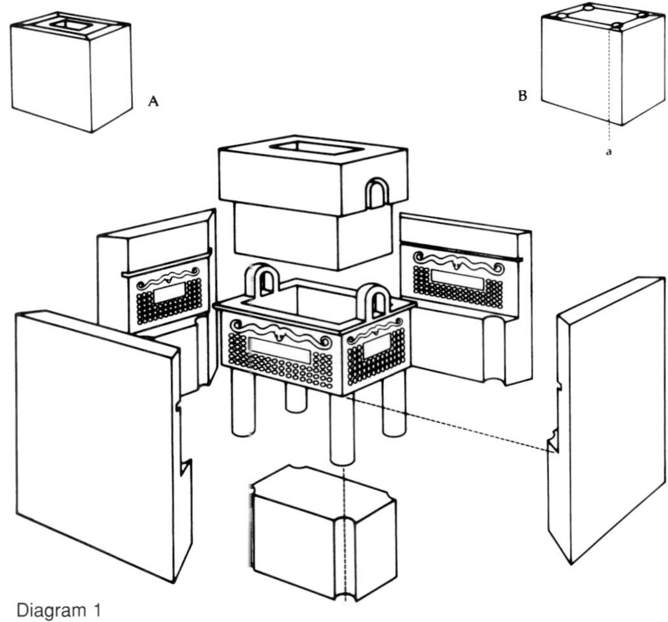
Diagram 1

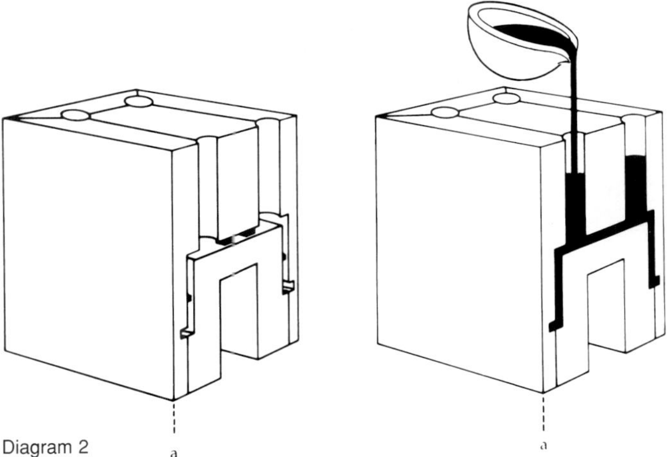
Diagram 2

Diagram 1. Mold assembly for casting a *fangding*. The model for the *fangding* (center) is surrounded by four panels comprising the mold. The inner core and interleg core sections are shown above and below the model. (A) The mold is shown assembled without the model. (B) The mold has been inverted for molten metal to be poured into a leg opening; (a) section removed in Diagram 2.

Diagram 2. Cutaway section of mold assembly showing casting of the *fangding*. The inner and interleg cores are kept apart by chaplets depicted as black squares (left). Molten metal is poured in through the leg opening to create the *fangding* (right).

Diagrams 1–2 reprinted from *Ancient Chinese Bronze Art*, drawings by Peter Lukic © 1991 China Institute of America; after an illustration by George Kelvin in *Emergence of Man: The Metalsmiths*, © 1974 Time-Life Books Inc.

access to the mold interior for retouching. Pyramid-shaped mortise and tenon joints insure proper alignment of the mold sections when assembled for casting.[6]

The assembled mold was then fired to harden it.[7] Prior to casting, the mold was disassembled and coated, either by smoking over a smudgy fire or painting with a mixture of carbon (lampblack or charcoal) in water.[8] The coating facilitated metal flow into the mold by reducing the surface tension of the molten metal against the mold surface.[9] It also acted as a separating agent to aid in removal of the bronze from the mold.[10] Evidence of the applied mold dressing sometimes can be seen in the bottoms of vessels as a layer of black between the foot and core remnants, and on mold fragments.[11]

As the mold defines the shape of the bronze exterior, cores are used to define the interior. Cores are solid forms made of loess that occupy the interior spaces of the mold. The gap left between the mold and the core is the casting space to be filled by the molten metal that creates the bronze vessel.[12] The core material is porous, which allows for the escape of gases generated during casting, to reduce casting failure. Cores could be made by shaving down the model to create the casting space.[13] Loss of the decorated surface on the model in making the core would mean that each bronze made in this manner was a unique creation. Cores not made from models were probably baked, but less so than the mold, to impart some strength yet be weak enough to break under pressure as the molten bronze shrank during cooling and to facilitate removal of the core after the vessel was cast.

Cores could also be used in vessel handles, feet, and other appendages to make them hollow rather than solid so they would be less susceptible to shrinkage and other forms of metal stress and failure.[14] These cores might either touch the outer mold to leave a hole in the metal, or contain core extensions where an insert made of core material bridges the gap between core and mold. The core extensions register the position of the core inside the appendage so it stays centered and maintains the casting space. Both the openings and core extensions would also allow gases to escape during casting. All of the appendages containing cores in bronzes examined for this study have some type of opening, which seems to indicate that they are necessary for successful casting, to vent gases or for registration. In addition to those having loess cores, appendages such as legs, handles, and knobs may be solid or contain copper cores.[15]

All authentic vessels are believed to have chaplets, or spacers. These are squarish bits of metal that separate the core from the mold to maintain the casting space and are incorporated into the bronze vessel in the casting operation.[16] They could be applied to the cores, possibly with clay slip to hold them in place.

Once the cores, appendages, chaplets, and mold sections were complete the molds would be assembled. Handles and other appendages were either cast as one with the vessel or cast separately and set in the mold prior to the final casting. The prepared mold could be bound with bamboo or ropes and buried for casting; several molds could have been done at the same time. Typically, vessels were cast upside down so the legs or base of the object would serve as pouring inlets for the molten bronze.[17] It is not known if the molds were heated prior to casting.[18] The molten bronze would have been heated in a crucible, or tappable furnace. It was poured from the crucible, ladled, or allowed to flow by gravity through a trough into the molds.[19]

The as-cast surface appearance of the bronze is thought to be shiny with a slightly frosted appearance as if it had been sandblasted, in a likeness of the fine grain of the loess comprising the mold.[20] The casting would contain rough seam lines called "mold marks" where metal flowed into the mold joints.[21] These would be removed through the use of gritty abrasive materials such as sand. Mold marks can be found on areas of the bronze that were either difficult to polish or areas where they would not be noticed, such as recessed areas of decorations, or the undersides of flanges. Occasionally mold marks are visible where the molds slipped or were out of plane, resulting in misaligned design elements or a metal step or ridge.

Natural stones and grits were used not only to remove mold marks, but also to polish the surface of the bronzes. Characteristically, the earliest bronzes of the Shang dynasty show the roughest surface finish, with ever-increasing degrees of polish in the Western to Eastern Zhou dynasties.[22] It is thought that at least some bronzes would have been polished to show the color of the metal alloy without further change.[23] The color of the "naked" bronze would range from orange to yellow to white with increasing amounts of tin. The addition of lead reduces the intensity of the color.[24]

The surface appearance of ancient Chinese bronzes as they exist today is a combination of what can be an original surface finish and alteration products (corrosion) that have occurred during burial, as well as mechanical and chemical processes that have been performed by dealers, restorers, and conservators after excavation. A thorough examination of the bronze using a variety of visual, technical, and scientific apparatus can reveal what has occurred over time.

In looking at the range of corrosion products on bronzes, one can see that generally "the mechanism of corrosion varies considerably from piece to piece and even from place to place within a single vessel. This results from factors in the local environment, such as the moisture content of the soil, salts, organic matter, bacterial action, permeability to atmospheric oxygen, and contact with other metals" and materials.[25] The latest theories indicate corrosion mechanisms may be driven by electrochemical processes in which ion migration, transport, and movement occur.[26]

The amounts of corrosion are also influenced by the degree of surface finish and alloy composition. A well finished surface can slow the rate of corrosion and affect the corrosion products formed. Tin added to the bronze makes the metal more brittle and capable of taking a higher polish, rendering it more corrosion-resistant. The combination of the extent of surface finish and alloy composition can determine the type of corrosion that occurs in the form of tin oxides on the surface, which in turn influences the protective nature of the corrosion film formed.

Keeping the factors influencing corrosion processes in mind, it is thought that during burial bronzes undergo two main types of corrosion processes. They have been described in the literature as Type I (delta removal) and Type II (alpha removal).[27] Delta and alpha refer to the different phases or crystal forms that appear in the bronze microstructure. They are selectively attacked in a corrosive environment depending on the nature of the metal, surface treatments, and burial conditions.

Type I (delta removal) corrosion products are the ones commonly seen on Chinese bronzes. They are mineral alteration products of the bronze constituents, mostly copper (cuprite, malachite, and azurite) as well as tin (cassiterite) and lead (cerussite) corrosion products.[28] They can appear as smooth or crusty patches of red, green, blue, gray, or white material. They occupy more space than the metal from which they are derived to create thick disfiguring crusts over the original bronze, at times disguising decorative features. The crusts are not cohesive, so they allow harmful products such as chlorides found in salts from the burial environment to migrate deeply into the bronze.[29] The chlorides react with the metal crystalline structure to promote active corrosion (bronze disease) even after the bronze has been excavated. It has been proposed that these less stable, thick, disfiguring corrosion layers are more likely to form on low-tin bronzes (which have a higher copper content), and those that were not originally patinated in antiquity whose unprotected "naked" bronze surfaces are more receptive to corrosion.[30] The Type I patina is considered less stable and not as visually appealing as the Type II patina.

Type II (alpha removal) corrosion products are associated with a smooth compact gray-green eggshell-like surface on the bronze, sometimes called a "water patina."[31]

It is mainly a thin layer of hydrous tin oxide that replaces the copper constituent of the bronze which has leached out of the alloy without a change in volume. The greenish color is due to staining from copper residues. It is a compact cohesive layer which forms more on the metal surface than in the metal microstructure as seen in Type I corrosion.[32] These properties make it more impervious to subsequent corrosion. Cuprite is not usually present as in Type I, although malachite and azurite can be deposited on top of the water patina. The formation of the compact, more stable Type II patina is thought to occur more readily on high-tin bronzes (with a lower copper content), and on bronzes that might have been patinated in antiquity.[33] The Type II patina is considered more structurally stable and, depending on aesthetic tastes, more visually appealing than Type I.

A detailed analysis was done on ten of the Chinese bronzes in the collection, involving conventional X-radiography, lead isotope analysis, and, in some instances, X-ray fluorescence and X-ray powder diffraction. On several bronzes the X-radiography revealed some interesting features that could not easily be discerned due to limitations of the technique. The decision was made to examine five of the bronzes using X-ray computed tomography (CT) to obtain a better look at the interior structures of the bronzes. Tomography work was done at GE Aircraft Engines Quality Technology Center in Cincinnati, Ohio. Since this is a relatively new procedure in the conservation field, a brief description of the method will be outlined.[34]

Industrial CT is a process similar to its more familiar medical equivalent, computerized assisted tomography (CAT) scanners. The main difference is that CAT scanners have been developed to provide high-quality cross-sectional images of the human body, but they cannot scan large dense objects. Industrial CT systems are designed to handle dense materials and larger objects using higher-energy X-ray sources. In industry, the procedure is typically used to inspect metal components ranging from large rocket motors to small precision castings in aircraft engines.

The system used at GE Aircraft Engines Quality Technology Center is capable of examining objects 36 inches high by 72 inches in diameter using either a 420KeV or a 2MeV X-ray source. The object is placed on a turntable in a lead-lined room. The system contains an X-ray source that generates a fan-shaped beam on one side of the object, with a corresponding row of detectors that receive the beam signal on the opposite side. The thickness of the beam determines the thickness of the cross section. The X-rays pass through the object as it is rotated on a platform and are picked up by the bank of detectors on the opposite side. The X-ray signal passing through the object is read from each detector and is converted to a numeric value. The values are then processed by a computer to create the CT image. In essence, a two-dimensional image is reconstructed from a set of one-dimensional radiation measurements taken at different scanning angles. As the rotating object is scanned, the X-ray source and detectors can be moved up and down at varying speeds.

This can produce three types of images: a computerized tomography slice (CT slice); a digital radiograph (DR); and a model. All of these images appear on a computer screen and can be stored on magnetic tape or computer disc for subsequent retrieval. The resulting image data can be manipulated in the computer in a variety of ways. Images can be enlarged or reduced, or further defined by adding color or contrast. In the case of a model, any number of cross sections of the object can be taken at any angle. Images can be reproduced in the form of laser copies and photographs. Sophisticated computer systems at the Quality Technology Center can essentially rebuild a model of the bronze in any number of ways, including from the bottom up, or the inside out. These images can be recorded in video format.

In conducting our examination, each of the five bronzes studied was quickly scanned overall to create a digital radiograph (DR) in a matter of seconds. The DR looks much like a conventional X-radiograph. It can reveal many features of the bronze, including areas of previous repair and locations of chaplets and core material. The DR was used to mark locations for CT slices through the bronze. The CT slice is essentially a slice through the object. It is the same principle as cutting through a a stick of pepperoni to obtain a slice. You can cut anywhere along the length of the sausage to obtain a slice that shows what the interior looks like. The same holds true for a bronze. The CT slice can, among other things, reveal porosity in the bronze, core material, inclusions of other materials, core extensions, repairs, and mechanisms for attaching component parts. This type of information is a great advantage over conventional X-radiography where the X-radiograph not only distorts the image produced on film, but also shows features of the bronze stacked on top of one another. The CT slice can allow for precise measurement of dimensions and enables one to single out a feature of the bronze for study without interference from surrounding structures.

Once the DR and CT slices were taken, some of the bronzes were scanned in detail to create computer models. This is a more time-consuming procedure that can take anywhere from a few hours for a portion of a bronze to overnight for a complete scan. In essence, the entire object in the case of *fanglei* No. 24 (2:1941), or portions of the object as in *jia* No. 20 (221:1950) were scanned by taking numerous CT slices in 1 mm increments. The resulting stack of CT slice data was combined to mathematically re-create the vessel, or a portion of it, in three dimensions. The computer-generated model could then be dissected in any direction, thereby providing access to interior and hitherto-unexplored regions of the bronze.

Gu
No. 1 (29:1951)

The *gu* was cast in a two-part mold assembly with vertical divisions on the sides, which bisect the open crosses. Mold marks have been incorporated into the thread relief decoration as vertical lines on the foot. The thread relief decoration on the foot was probably drawn in the mold without the use of a model. The two bands of rings encircling the *gu* could have been stamped with a bamboo tube.[35] A mold mark is visible as a dark line in the patina on the outside of the body. Others are present in the wall thickness inside the two open crosses on the foot, indicating there were core extensions for them in the foot core as well as the mold. The mold mark is slightly offset in each cross, showing an inexact alignment of the inner and outer core extensions (Figs. 1.1–2).

The *gu* exhibits a smooth compact water patina that is mottled from light to dark olive green. The dendritic structure of the metal has been preserved and can be seen under 20x magnification. The water patina contains a fine crackle network. It has actively flaked off in some areas to reveal an underlying granular pale green surface, with brown bronze beneath.[36] Most of the loss has occurred at the rim. Losses were filled and overpainted in brown.

Interior areas of the vessel neck and foot are encrusted with patches of malachite, azurite, cuprite, and dirt. Surfaces have been mechanically cleaned to remove encrustations down to a water patina, probably to improve the vessel's appearance. Their mechanical removal resulted in numerous surface pits and craters with

Figure 1.1. Detail of a cross opening. A mold mark appears as a vertical line (magnified approximately 5x). Cross-shaped extensions were present in both the foot core and the mold.

Figure 1.2. Detail of the opposite cross opening on the foot, magnified approximately 5x. Note the misalignment of the inner and outer portions of the cross due to inexact registration between the cross-shaped extensions in the foot core and the mold during casting.

Fig. 1.1

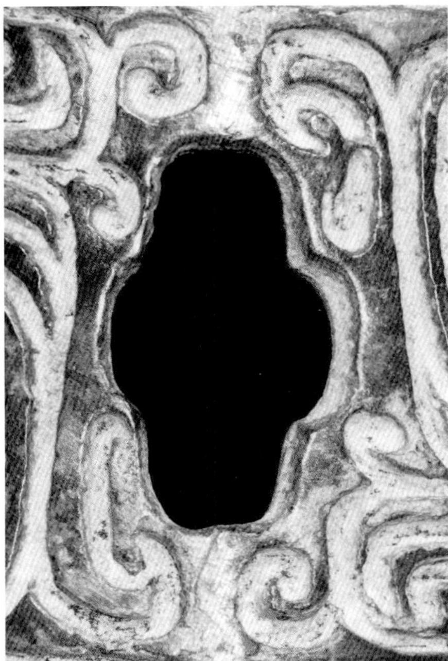

Fig. 1.2

exposed brown bronze beneath. They have been subsequently covered with gritty paints to simulate archaeological corrosion. Painted areas fluoresce bright yellow and dull orange under long-wave ultraviolet light.

The thread relief design on the foot contains numerous scrape marks from mechanical cleaning, especially in incised areas. Losses in the relief decoration have been filled and overpainted. An intaglio design of a doglike animal is present on the interior foot; it is not original to the *gu*. The tip of the tail was mechanically cleaned to gain a better understanding of the surface modification. The animal has been overpainted to simulate archaeological corrosion and fluoresces bright yellow under long-wave ultraviolet light. It was created after the *gu* was made by scraping away the original corrosion crust to obtain a smooth work surface in the general shape of the animal. Removal of the restoration showed that the smooth scraped area was covered with blue-green paint overlying a metallic brown surface where the animal decoration has been gouged out with a tool. The gouged edges have been built up in relief with a brown waxy material to simulate a metal edge. The waxy material has a broken, beaded appearance as if drizzled on the surface, instead of being a continuous raised line that was cast.[37] The raised edges do not appear dense in X-radiographs as they would if they were made of metal (Figs. 1.3–4).

X-radiography revealed a chaplet in the bottom of the *gu* at center. Two dark spots on the sides of the *gu* may be two more chaplets that have drifted out of position. There is not much porosity in the bronze. Where present, it is more concentrated in the body than the foot, suggesting that the vessel was probably cast right side up.

Fig. 1.3

Figure 1.3. Detail of the animal inscription magnified approximately 5x. Crusty corrosion products surrounding the animal have been scraped smooth. The animal shape has been gouged out of the metal and outlined with a beaded ridge of dark wax. The entire area has been painted to simulate corrosion.

Figure 1.4. The tip of the animal's tail with wax and paint removed to reveal the scraped surface and toolmarks where the tail was gouged out. Magnified approximately 8x.

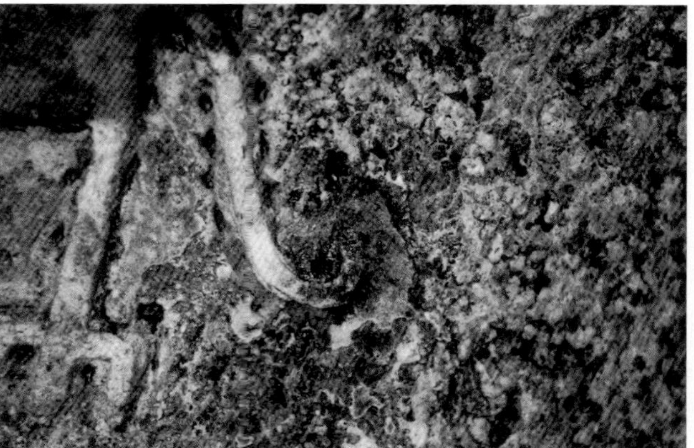

Fig. 1.4

Gu
No. 8 (217:1950)

The *gu* was cast in a two-part mold assembly; mold divisions are located down the center of the rising blades on the sides that do not contain the crosses. Mold marks are visible at the top of the bowstrings on the rising blades where the incised decoration is slightly misaligned (Fig. 2.1). Elsewhere the alignment of decoration along mold divisions is extraordinarily fine. The mold contained two interior core pieces, one for the body and one for the foot. They have been separated by a large chaplet now incorporated into the septum.

Crosses are present on the foot of the *gu* directly below the septum (Fig. 2.2). Aside from being decorative, their function is to register the core.[38] They are partially filled in, to roughly half their depth; their depressed outlines show on the exterior surface. The partial filling is due to incomplete contact between the core inside the foot and the cross-shaped core extensions in the mold which allowed metal to flow in and semi-fill the crosses during casting. The location of the cross depressions on the exterior surface implies the core extensions were part of the mold rather than the core.

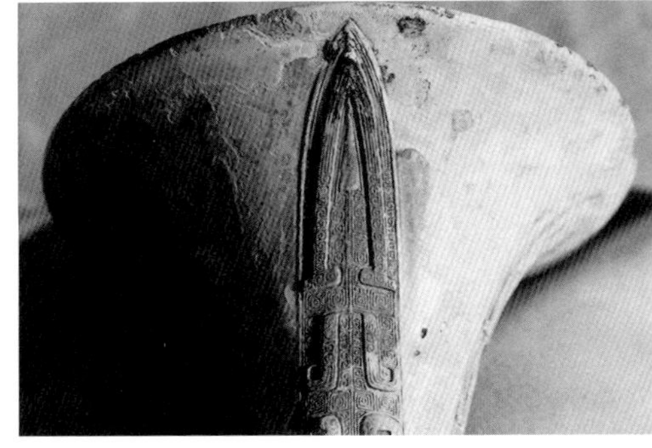

Fig. 2.1

Residual core material may be present inside the foot as encrustations of granular red-orange products around the edge of the septum. The interior of the foot contains particles of copper corrosion products (cuprite and malachite) and carbon. The interior bottom of the foot contains four longitudinal buttresses (Fig. 2.3). They correspond to the plain ribbed band containing the crosses on the exterior. Their function is to prevent hot tearing of the metal as it cooled.[39] Buttresses do not appear in all *gu*. They may be indicative of *gu* from Anyang, although this would require further research. There are ridges of mold flash along the inside wall from cracks in the foot core. There are black residues of carbonaceous material that ring the inside of the foot, mainly at the top and bottom. There is a cuprite pseudomorph of a charcoal fragment near the bottom edge of the foot.[40]

The overall tin oxide water patina is mottled yellow-green with smaller patches of dark olive and black sulfide products on top. There are large patches of gray and turquoise corrosion products on the inside mouth and neck of the vessel and on the exterior mouth. Both the yellow-green patina and the blue-gray one occur at the same corrosion level on the bronze. The color differences may be due in part to different oxidation states of the tin oxide and staining from copper corrosion

Figure 2.1. Detail of the tip of the rising blade decoration showing the slight surface protrusion of the mold seam line where there is a misalignment of the decoration.

Figure 2.2. Partially filled cross in the foot area. The partial metal filling is due to incomplete contact between the interior foot core and outer cross-shaped extensions that were part of the mold.

Fig. 2.2

products. There are intermittent warts of cuprite and malachite on the vessel exterior.

The dendritic structure of the casting is visible on the yellow-green and the gray-turquoise areas of patina. The relatively large size of the dendrites is an indication of slow cooling of the metal after casting. In the metal structure the alpha portion of the metal has corroded with the delta phase left behind.[41] This is characteristic of a Type II patina.[42]

The interior neck of the *gu* is stepped in above the flanges, possibly to maintain an even wall thickness. The same inverted feature can be seen on *zun* No. 14 (125:1951). The function of this step may be to prevent hot tearing of the metal during casting (Fig. 2.4).

A chaplet measuring roughly two cm square is highly visible in the septum. Viewed from the bottom, it is slightly recessed. It is covered with red-orange redeposited copper, dark red cuprite, and one blister of malachite (Fig. 2.3).

X-radiographs of the object revealed minimal porosity in the metal, an indication of fine casting.

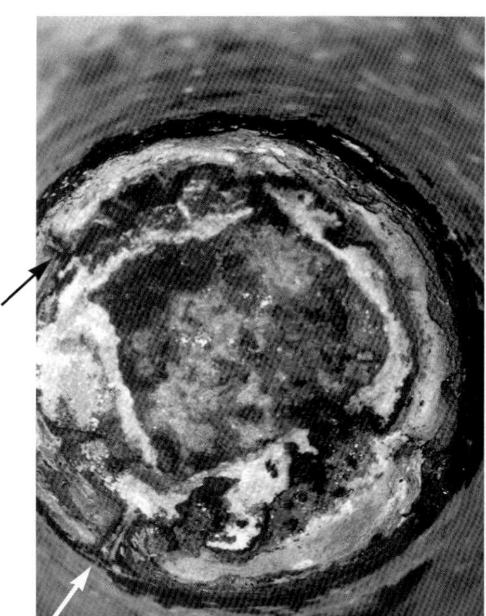

Figure 2.3. Interior foot magnified approximately 8x. Two of the four buttresses are visible as short ridges (arrows). A large chaplet is apparent as a dark square in the center of the foot.

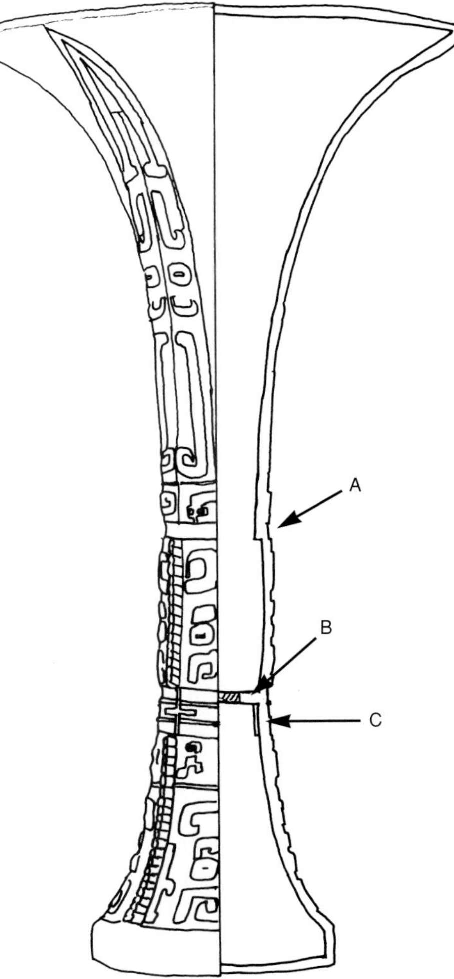

Figure 2.4. Sketch of the *gu*. The right half has been cut away to reveal an estimated profile of the vessel walls: (A) the step in the throat; (B) a portion of the chaplet; (C) a buttress.

Fig. 2.3

Fig. 2.4

Zun
No. 14 (125:1951)

Figure 3.1. A view of the bottom shows buttresses where the septum meets the vessel wall (arrows).

Figure 3.2. Detail of the midsection, showing the region where the false patina has been removed.

Figure 3.3. Details of the cast intaglio decoration magnified approximately 13x. A false patina of green paint has been removed to reveal an underlying pitted metal surface.

The object was cast in a four-part mold assembly that divided at the flanges. Mold marks are present on the bottoms of three of the lower flanges where seams are misaligned. The *zun* contains casting features similar to *gu* No. 8 (217:1950). Both vessels have the same general shape. The interior bottoms of the feet on both vessels contain four longitudinal buttresses. On each they correspond to the plain ribbed bands running between the two sets of flanges on the exterior. Likewise, the interior necks of the *zun* and *gu* are stepped in above the exterior flanges, probably to maintain an even wall thickness. The probable function of the buttresses and the steps is to prevent hot tearing of the metal as it cooled (Fig. 3.1).[43]

A detailed examination of the *zun* raised many interesting questions about its current state. Most of the vessel has been painted to simulate corrosion. Beige dirt as well as cuprite and malachite accretions have been applied to appear as natural corrosion formed during burial. The beige dirt in particular fluoresces yellow-green under ultraviolet light. The bronze has been painted to imitate a smooth, semigloss, mottled light and dark green water patina. Mud-colored brown dirt has been painted in the inscription and incised decoration. These paints can be removed with solvents. Overall, there is no evidence of genuine archaeological corrosion penetrating the metal; instead, the metal surface is pitted and irregular, similar in appearance to metal that has been cleaned with acid (Figs. 3.2–3).

Taotie relief decorations occur on the middle and lower registers of the vessel. *Taotie* features such as eyebrows, eyes, noses, cheeks, and mouths have been selectively incised with narrow lines that have been cast in, then filled with paint. There is an intaglio inscription on the interior bottom of the vessel. It is also covered with brown paint, and there is no evidence of archaeological corrosion.

X-radiography revealed that two chaplets are present on either side of the inscription at midsection. The body of the vessel is somewhat more porous than the foot, though it is unclear as to whether the object was cast right side up or upside down.

The nature of the intaglio lines, metal surface, lack of archaeological corrosion, and applied paint, along with lead isotope ratios, raise questions about the vessel. Further study is needed in order to characterize it accurately.

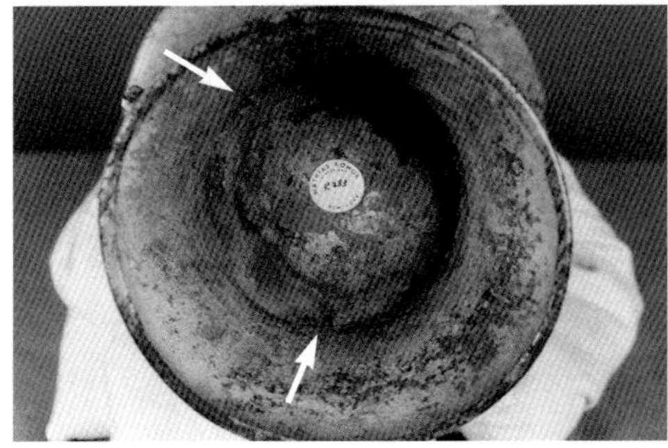

Fig. 3.1

Fig. 3.2

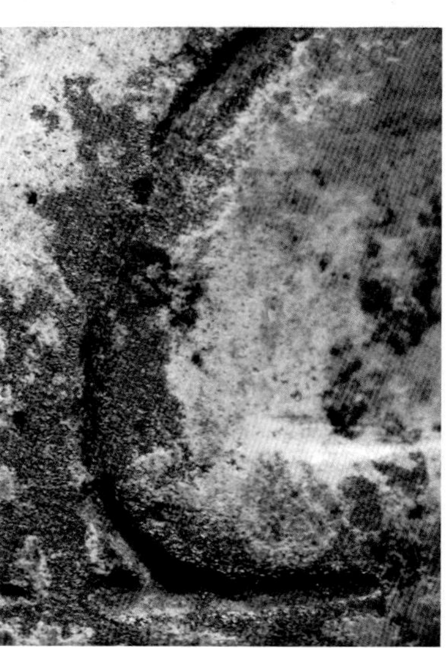

Fig. 3.3

Fangding
No. 15 (108:1954)

The *fangding* was cast in a four-part mold assembly with divisions at the corners (see Diagram 1, p. 149). Mold marks are visible on the underside of the corners at the rim and on the bottoms of the flanges. The legs contain mold marks from an interleg core piece which appear as faint relief lines behind the horns of the *taotie* and at the junction between the vessel bottom and one leg. The legs contain rectangular core extensions on the interior at midsection (Fig. 4.9).

Surfaces are decorated in intaglio and relief patterns. Some relief decoration would have been done in the mold such as the raised bosses with and without protrusions surrounding the snakes. The protrusions taper from base to tip, a probable indication that they were made by pressing a blunt-pointed tool into the mold. The intaglio decorations were most likely done on the model. Evidence suggests there were problems in transferring the designs from model to mold: incised lines on the handles are particularly shallow and indistinct; and the T-hook design, *lei-wen*, and designs on the snakes all contain areas that have been flattened. It appears they were flattened against the model while it was still in a plastic state when the designs were being transferred from model to mold. The greatest degree of flattening is located at the center of a

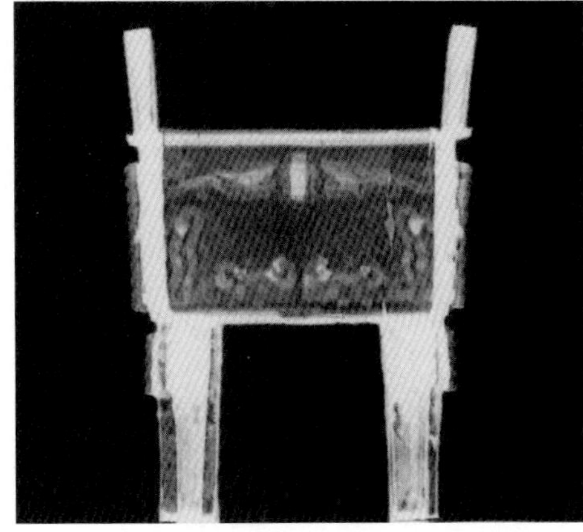

Fig. 4.1

side panel where the inscription occurs (Fig. 4.1). Here the external design features have not only been flattened, but also pushed out of plane. The distortion appears to have been produced by the pressure used to stamp the inscription into the model (Figs. 4.2–3). The intaglio lines of the inscription appear to be of uniform depth and their walls are straight (though they seem to taper to a shallow ending at the tops), as one could find in a stamped design.[44] A circular 0.5 cm loss of design is present in the chevron in the lower right corner of the inscription, apparently due to a flaw in the mold during transfer of the design as well as injudicious cleaning.

Exterior surfaces are mainly covered with a green-gray and glossy tin oxide water patina. Glossy surfaces are covered by scratches that are visible to the unaided eye in strong light. They appear to be from original finishing operations, as they can be found under archaeological corrosion and black products (Fig. 4.4).[45] There are intermittent crusty patches of malachite, cuprite, and azurite. Many of the

Figure 4.1. DR of the *fangding*. The intaglio inscription at top center appears dark. Solder repairs on the right side appear white.

Figure 4.2. Detail of side panel, magnified approximately 3x. The T-hook pattern has been distorted by flattening, possibly as a result of the inscription having been stamped in behind it.

Figure 4.3. Verso of the side panel, containing the intaglio inscription, magnified approximately 3x. The plain circular area is probably the result of injudicious cleaning of the warty corrosion.

Figure 4.4. Rim area of the *fangding*. The reticulated black product may be a coating, possibly a lacquer that was applied in antiquity. Note that polishing scratches continue from the black to the underlying lighter region at left, indicating they are part of the original finishing operations.

Fig. 4.2

Fig. 4.3

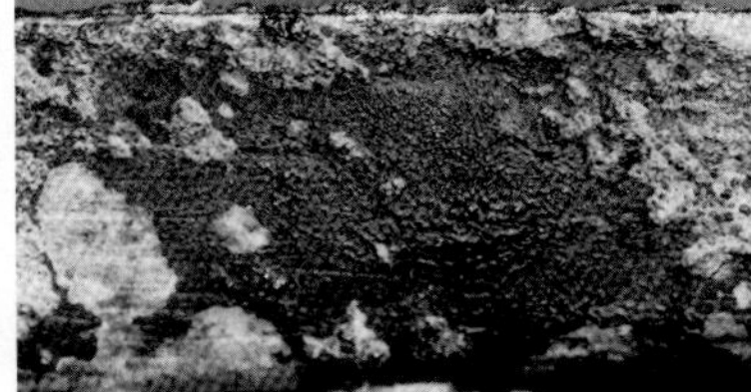

Fig. 4.4

intaglio designs are filled with black products, as are portions of the rim and bottom. These may be deposits from use over a fire, or intentionally applied material.[46] Rectangular core extensions on the legs are slightly recessed and filled with crusty light green and olive corrosion products; their function was to register the cores and possibly to permit the escape of gas during casting.

Remains of fabric pseudomorphs occurring on the rim above the inscription are similar in location and appearance to those on *fangding* No. 16 (222:1950), leading one to suppose the inscription in particular was protected by a fabric covering. There are also pseudomorphs of at least two different types of fibers preserved in the malachite corrosion on the interior.[47] In general, the interior is heavily encrusted with dirt, malachite, and white products possibly from lead repairs. Some reticulated black residues are present on the sides near the rim. The bottoms of the feet contain reedlike pseudomorph impressions where the vessel sat on some type of mat (Fig. 4.5).

X-radiographs reveal the legs and handles were cast integrally with the vessel. The legs are hollow and contain cores, and the handles are solid. Porosity at the base of the handle indicates the vessel was probably cast feet-up. Two, possibly three,

Fig. 4.5

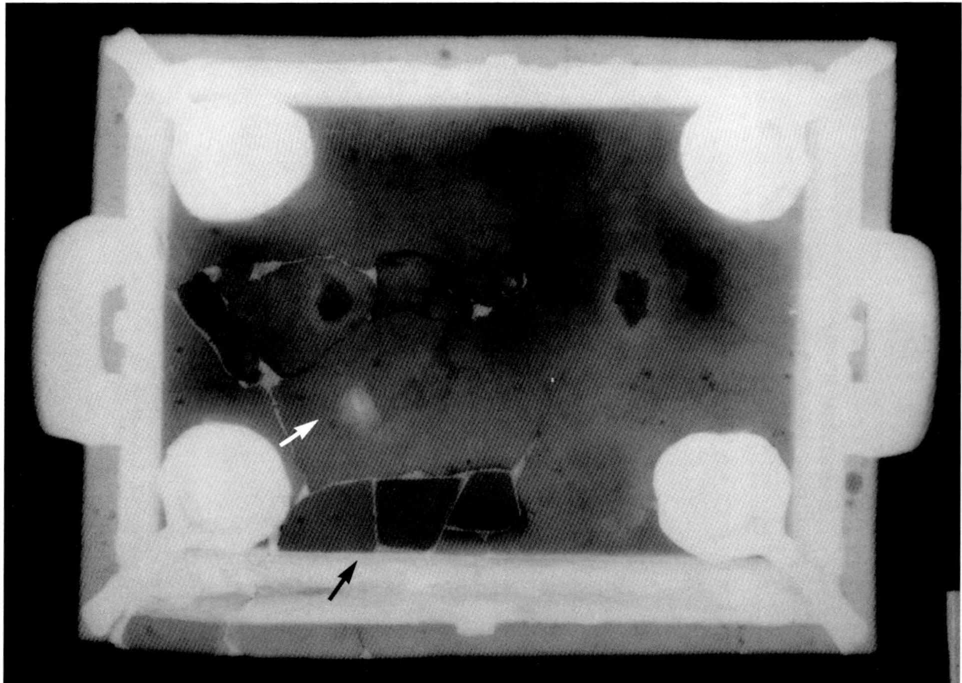

Fig. 4.6

160

Figure 4.7. A CT slice through one side and two legs. The repaired leg contains a support rod running through the center. The lower third of it has been replaced with modern metal. Note the uniform thickness of the metal walls. The original core material in both legs is light gray. The round white inclusion above the rod in the repaired leg may be metal filling a void in the core. Note that the leg was at an angle when the slice was taken so that not all of the rod is visible.

Figure 4.8. View of the bottom. The lower left quadrant contains a plug of metal to fill a hole where a chaplet probably fell out. The plug contains a residual linear stump where the sprue was broken off after pouring.

Figure 4.9. A DR showing all four legs. The metal rod and replaced portion of the repaired leg is plainly visible. Rectangular core extensions appear as dark rectangles on the two center legs. Core material in the legs appears gray.

Figure 4.10. A CT slice through the core extensions in the legs. The core extensions appear as discrete rectangular inserts in the leg cores. The support rod appears as a white circle in the center of the repaired leg.

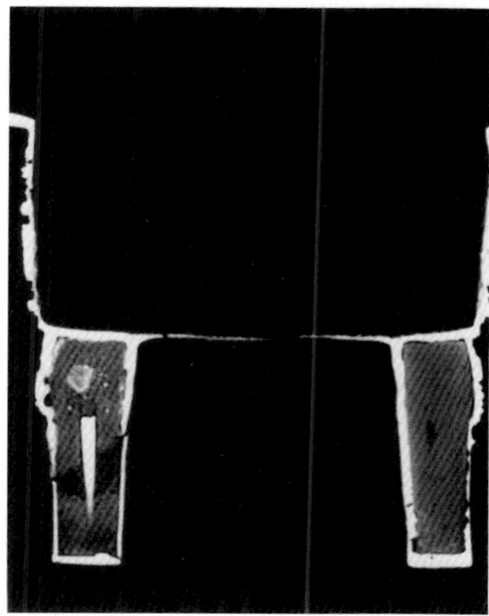

Fig. 4.7

chaplets are evident in the vessel bottom as well as an ancient repair where a chaplet fell out. The repair contains more lead than the surrounding bronze as it is radiopaque. The *fangding* has been broken and extensively repaired on the bottom and two sides with lead solder. Three pieces of modern metal have been inserted and soldered into the bottom (Fig. 4.6). Roughly the bottom third of one leg has been replaced and contains an internal support pin. The top portion of the leg, decorated with the mask, is original (Fig. 4.7).

Under visual inspection, the repaired areas are covered by murky paint mixed with corrosion products. They are indistinct and muddy compared to original archaeological encrustations. They fluoresce dull orange under long-wave ultraviolet light. Some areas on the vessel appear whitish, possibly from lead solder repairs. An ancient repair is present on the vessel bottom where a hole, probably from a lost chaplet, has been filled with a patch of metal. It contains a residual stump where the sprue was broken off after pouring (Fig. 4.8).[48]

A DR of the legs clearly shows the replacement leg and support pin are of modern metal as the metal is of uniform thickness in comparison to the ancient bronze (Fig. 4.9). A CT slice through the legs reveals that the rectangular core extensions are separate inserts placed in the leg cores (Fig. 4.10), and probably formed part of the interleg core piece.

Fig. 4.8

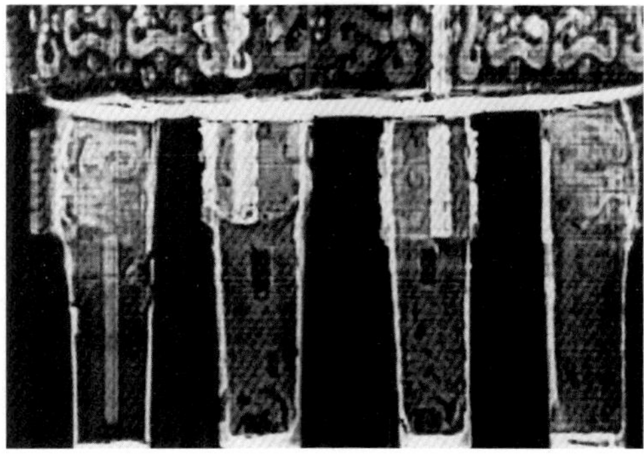

Fig. 4.9

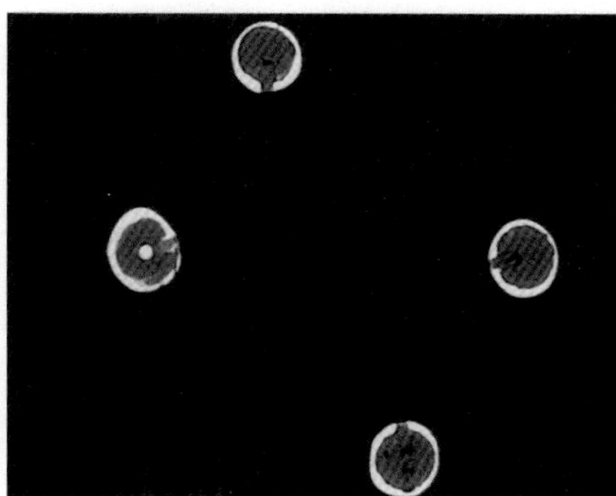

Fig. 4.10

Fangding
No. 16 (222:1950)

The *fangding* was cast in a four-part mold assembly with divisions at the corners (see Diagram 1, p. 149). Mold marks are not discernible. Handles and legs were cast integrally with the vessel. The vessel is decorated with plain center panels surrounded by protruding hobnails on the sides and a band of incised decoration near the rim and on the legs. The incised designs were most likely done on a model while the relief decorations such as the rows of hobnails, *taotie* eyes, and flanges were probably done in the mold. Gray-colored rectangular core extensions are visible on the inner mid-sections of the legs where the leg cores contacted the interleg core piece.[49]

The interior is encrusted with thick layers of cuprite, malachite, azurite, and dirt. They have been removed from the area around the inscription and from most of the vessel exterior to expose a smooth, compact, mottled green-gray tin oxide water patina. Polishing scratches are visible on the water patina. Exterior surfaces contain patches of smooth, somewhat glossy, fine-grained black products on top of the water patina, which are also intermingled with residual cuprite and malachite crusts; they may be deposits from use over a fire. Remnants of black material also occur in the incised decoration near the rim and completely fill the decoration in the legs as if intentionally applied.[50] In some instances, black lines have been painted on the legs by a restorer to complete the appearance of the intaglio lines where they are disfigured by large red-orange warts of copper corrosion products. Samples of black material were removed from the incised *taotie* decorations near the rim for SEM and FTIR analysis to determine whether it occurred naturally, or whether it was an intentionally applied material such as lacquer.[51] SEM analysis confirmed the presence of elements comprising the alloy as well as those associated with a burial environment.[52] FTIR results were promising in characterizing the sample as organic, relating it to rhus constituents in lacquer, though a direct correlation between the sample spectra and known lacquer spectra was not found.[53] Further research is planned to determine whether the material can be characterized as lacquer inlay.

The interior and rim portions of the vessel contain textile remains and pseudomorphs. There are two yellow-brown patches of partially mineralized fibers located on the rim and the midsection and on one corner (Fig. 5.1). The fibers on the rim edge appear to be bast fibers, possibly linen or hemp.[54] The mineralized fibers comprising the textile fragment on the rim midsection appear to be a plain weave fabric with a Z-twist.[55] Both patches may be part of a fabric that was draped over the inscription, or formed part of a wrapping for the vessel. The interior of the vessel is encrusted with strands and fragments of fiber pseudomorphs belonging to as many as four different types of yarns according to their size and appearance.[56]

The dendritic structure of the casting has been converted to tin-oxide corrosion products and is visible under low magnification. The relatively large size of the dendrites indicates slow cooling of the metal. Dendrite formations confirm that the inscription was cast in.

X-radiographs show cracks and breaks in the bottom and sides of the vessel. They have been repaired by tacking them together with solder. One leg has been reattached to the vessel bottom with solder. Four chaplets are present in the vessel

Figure 5.1. Detail of the rim magnified approximately 3x. Remnants of a plain-weave textile and textile pseudomorphs indicate the vessel could have been draped with cloth.

Fig. 5.1

bottom. They look dark in the X-radiograph and breakage lines connect three of them. All the legs are hollow and contain cores which are out of alignment. Dark rectangular core extensions are visible in the midsection of the legs. The handles are solid metal (Fig. 5.2).

The solder repairs have been overpainted with bits of malachite to look like naturally formed corrosion products. They are slightly darker in color than surrounding areas. The bottoms of the feet have been covered with a chalky/waxy fill to make them even. The repairs fluoresce brightly under long-wave ultraviolet light.

Figure 5.2. X-radiograph taken at an angle. Thicker metal areas such as the vessel rim, hobnails, and solid handles appear white. Because the metal wall is thinner, the intaglio inscription appears dark in the center of the front panel. Solder repairs of cracks in the vessel bottom appear as white lines due to lead in the solder. Leg cores appear slightly dark. The far left one is particularly misaligned. A core extension is visible as a dark rectangle in the center right leg.

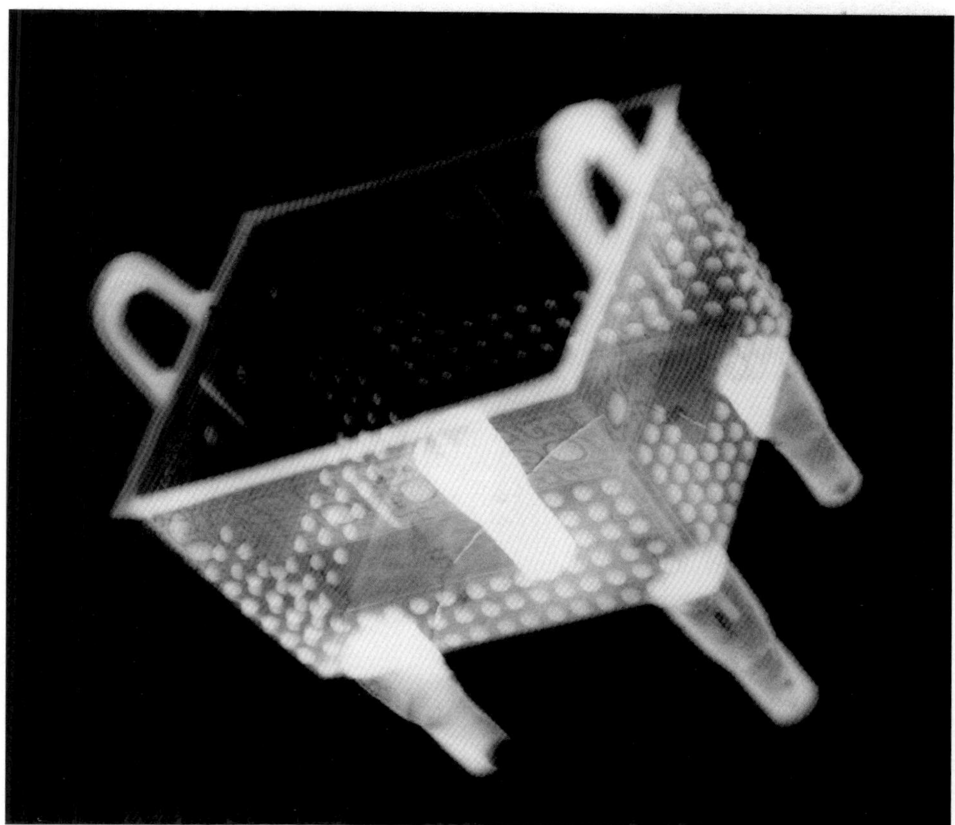

Fig. 5.2

Fangyi
No. 19 (127:1951)

The vessel was cast in a four-part mold assembly that divided vertically at the corners. Residual mold marks on the corners are indicated by faint raised areas beneath some flanges and relief elements. There is also some misalignment of the decorative band and relief line at one corner. Mold flash from the interior foot core is present on the inside of one of the inverted U–shaped openings on one short side of the vessel. Edges near the inverted-U opening on the opposite short side appear to have been flattened and pressed outward with resultant stress cracking. The damage appears to have occurred after the bronze was cast, as if the vessel had been struck against something. The rim and bottom edges of the vessel exhibit edge thickening that was probably done on the model prior to casting, to impart strength to both the model and cast vessel. The lid appears to have been cast in a four-part mold assembly that divided at the corners, as indicated by some misalignment in the decorative band. Mold marks are present beneath the knob and run in line with the peak of the lid. In general, the bands of intaglio decoration were most likely created on a model. Design flaws are present where metal has partially or completely filled the intaglio designs. These flaws could be the result of poor design transfer from model to mold. In effect, the mold did not receive the design, so it did not appear in the casting (Fig. 6.1).[57]

The surface of the *fangyi* is encrusted with velvety emerald green malachite marked with bubbles of botryoidal malachite, orange to ruby red cuprite, and dirt. There are fragments of a plain weave textile preserved as positive and negative pseudomorphs on the lid near the corners. The correct orientation of the lid on the vessel can be identified by matching corrosion patterns on both parts. The intaglio designs are encrusted with bright orange-red cuprite, malachite, and a black product. A sample of bright orange corrosion removed from a corner of the lid and analyzed by X-ray powder diffraction confirmed the material is cuprite.[58] The black product can be found not only in the intaglio designs, but also distributed over various areas of the vessel. It may be residue from the vessel's use over a fire, or an intentionally applied material (see FTIR information for *fangding* No. 16; 222:1950). Much of the vessel has been mechanically cleaned to reveal a glossy tin oxide water patina. It is mottled light and dark green with some tinges of gray, pink, and brown. Several areas have been cleaned down to a brown-black metallic layer and have subsequently been overpainted with gritty paints to simulate archaeological corrosion. The paint fluoresces bright yellow-white under long-wave ultraviolet light and can be removed with solvents. Areas around the vessel inscription have been cleaned to reveal the yellow-brown color of the bronze.

The dendritic structure of the metal is visible in the tin-oxide water patina under low magnification (13x). The relatively large size of the dendrites is an indication of slow cooling of the metal after casting. In the metal structure the alpha portion of the metal has corroded with the delta phase left behind, which is characteristic of a Type II patina.[59] X-radiography indicated the knob on the lid was cast integrally with the vessel and contains a core. Chaplets are visible as dark squares on the X-radiograph. Two were placed at the gable ends of the lid and four occur around the inscription in the vessel bottom. Some porosity is present in the side view of the vessel, though the casting directions could not be ascertained.

Figure 6.1. Top left band of decoration on the vessel, magnified approximately 8x. The unlined gray and white area indicates that the design translated only partially from model to mold prior to casting. Darker lines show corrosion in the original intaglio.

Jia
No. 20 (221:1950)

The *jia* was cast in a three-part mold assembly with vertical divisions bisecting the front of each leg. There would have been separate mold inserts for the posts, and a D-shaped one for the inside of the handle. Mold marks are present on the pairs of relief lines on the legs, and in the bands of decoration at the neck where there is some misalignment of design elements. A mold mark remains on the interior of the dragon's upper lip, identifying the location where the mold probably bisected the handle and corresponding leg. There are prominent mold marks left from the mold inserts on the underside of the handle and caps on the posts (Fig. 7.1). Beneath the caps they run parallel and perpendicular to the outer flat face of the posts. The bottom of the *jia* is outlined by triangular mold marks from an interleg core piece. The bottom is decorated with a series of Karlbeck lines in a chevron relief pattern where the interleg core was scored (Fig. 7.2). The possible function of these lines was to register the core and/or to allow for the escape of gas during casting.[60]

The underside of the handle contains a band of hard gray core material topped with patches of powdery white efflorescing material.[61] The midsection of the handle contains a lump of bronze inside the core material, which locks the core in place. It is not known if this was accidental or intentionally done in antiquity.

The outlines of three squarish chaplets are visible to the unaided eye on the underside of each leg at midsection. Smaller ones have been placed between the pair of relief lines decorating the legs.

Exterior surfaces are covered with a glossy smooth tin oxide patina that is mottled in shades of green to black. There are intermittent patches of cuprite in the interstices in the banded decoration at the neck and shoulder. Surfaces have been subjected to acid and mechanical cleaning post-excavation. Heavy scratch marks are present along edges of relief decoration where they have been wheel cut with a machine such as a Dremel tool to remove corrosion (Fig. 7.3). The acid treatment left portions of the surface pockmarked. These areas were cosmetically overpainted to

Fig. 7.1

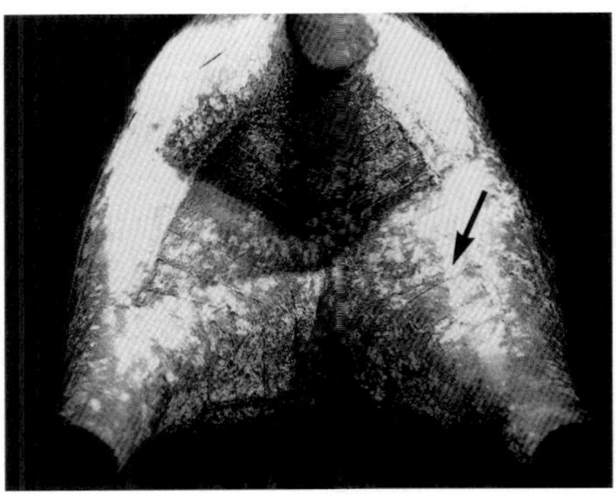

Fig. 7.2

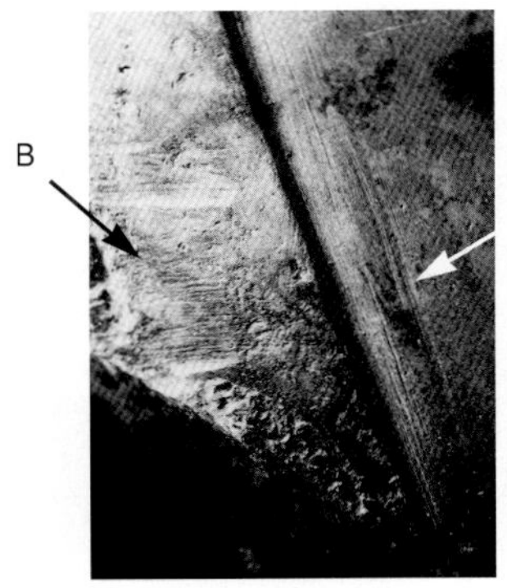

Fig.7.3

Figure 7.1. Underside of one of the capped posts. Mold marks run parallel and perpendicular.

Figure 7.2. View of the bottom. A series of mold marks called "Karlbeck lines" covers the entire bottom area.

Figure 7.3. Border of the inscription beneath the handle where it joins the vessel. Vertical scratches from a cutting wheel run parallel to the ridge in the center of the photo (A). Horizontal wheel-cut scratches are also evident above the handle (B).

Figure 7.4. A CT slice through the legs shows dark rectangular patches which may be copper cores in the solid portion of the legs.

Figure 7.5. DR of the *jia*. Notable features include the high degree of porosity in the casting. Note the extra inclusion of metal at the center of the handle. The feet are solid.

Figure 7.6. A horizontal CT slice through the bronze lump in the center of the handle.

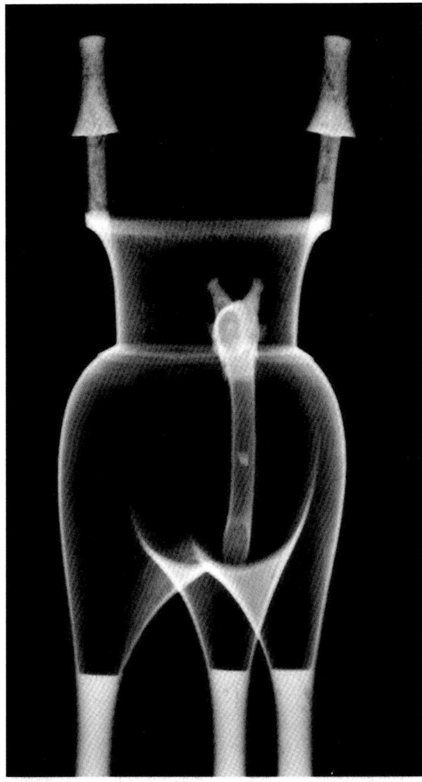

Fig. 7.5

imitate cuprite and bronze disease. They fluoresce bright yellow-green under long-wave ultraviolet light. The teeth on the dragon-head handle are dull gray in comparison to the rest of the bronze. They would have been a dramatic contrast to the yellow color of the bronze in antiquity. Visual examination, X-radiography, and tomography have not indicated whether they are a different alloy that was attached separately, or whether the color is due to a surface treatment. The latter seems likely. A lengthy delicate intaglio inscription was cast in beneath the handle. Ridges along the sides of the inscription have been enhanced by post-excavation wheel cutting. Its nearly inaccessible location beneath the handle suggests the inscription was inserted into the mold, though no seam lines are evident to indicate this.

The interior of the vessel appears matte and dry. A tideline divides it almost in half longitudinally, indicating the vessel was lying on its side during burial. One side is velvety green and accented by bright red cuprite spots. The other is rough and covered with tan-brown-gray tones with some green copper corrosion products. The three chaplets for the legs can be seen as squarish rusty-looking patches, probably cuprite.

X-radiography shows that the handle and posts and caps were cast integrally with the vessel. Chaplets in the legs appear as dark squares. There is a fair degree of porosity overall, but especially in the posts, which have a spongy appearance. The porosity seems to be rising from the vessel's feet up to the posts, indicating it was cast right side up.

Tomography of the *jia* revealed some interesting features, notably the presence of a different material in the legs which may be indicative of copper cores (Fig. 7.4).[62] Other features identified in the X-radiographs could also be discerned in the DR, such as the general porosity in the object, chaplets in the legs, the bronze lump in the core of the handle, and the solid legs (Fig. 7.5). A CT slice of the bronze lump in the central portion of the handle indicates a portion of the core in the handle was hollow, which allowed bronze to be poured in, possibly as a means of locking the core in place (Fig. 7.6).

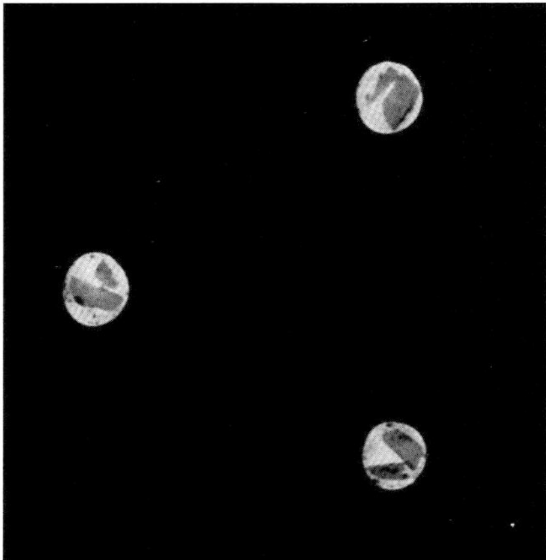

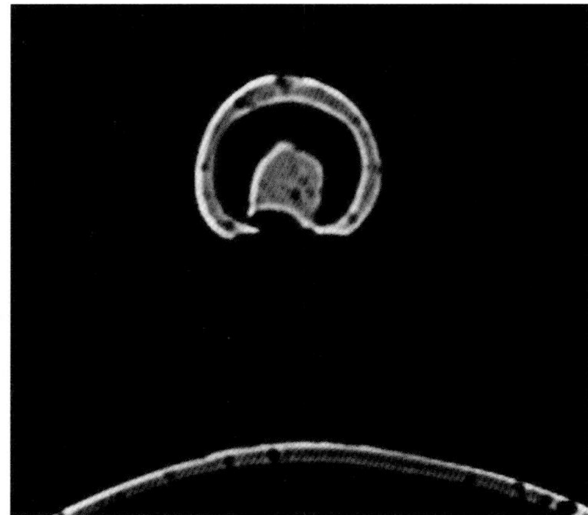

Fig. 7.4

Fig. 7.6

Zhi
No. 23 (215:1950)

The vessel was cast in a two-part mold assembly that vertically bisected the vessel at the ends of the long axis. Mold alignment seams are visible as step offsets on the decorated band on the foot at either end of the long axis. Chaplets are visible as darker squarish patches on the exterior undecorated band separating the upper portion of the vessel from the foot.

The vessel's outer surface is compact and somewhat glossy with a mottled dark and light green water patina. The exterior was probably mechanically cleaned to reveal the water patina by removing heavy disfiguring archaeological crusts of malachite, azurite, and cuprite like those found on the vessel's interior. The *leiwen* are encrusted with velvety red cuprite. There are a few remaining intermittent patches of cuprite and malachite scattered over the vessel, mainly around the foot of one of the decorative bird panels. Overall there are intermittent patches of lightly oxidized yellow-brown bronze metal showing through, especially at the midsection of the vessel in the bird's plumage. There are numerous patches of gray-green paint and a waxy fill material mainly in areas of higher relief where surfaces were heavily corroded or damaged in cleaning. The paint fluoresces bright yellow-orange under long-wave ultraviolet light.

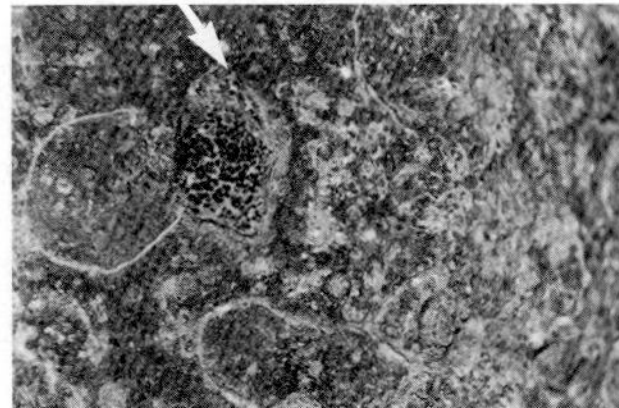

Fig. 8.1

The vessel interior is heavily encrusted with malachite, azurite, and cuprite archaeological corrosion products and dirt. The bottom crust is scratched from cleaning and cleaning attempts. There are circular rings left where botryroidal malachite bubbles have been broken off (Fig. 8.1). The archaeological crusts have been removed in three areas to reveal the underlying water patina, most likely in an attempt to find inscriptions.

The dendritic structure of the metal is visible under low (5x) magnification, mainly in the smooth water patina on the animal in the vessel bottom (Figs. 8.2–3).[63]

X-radiography revealed chips and cracks in the foot and rim. A repair in the lip is radiopaque; since X-rays cannot easily penetrate lead or barium, the white opacity suggests a fill of either of those materials, probably lead. Two locations where the foot was drilled for metal analysis appear dark.[64] They have been filled with pigmented wax.

Samples of an emerald green applied corrosion product were taken from the rim and analyzed by X-ray powder diffraction.[65] It was identified as Paris green paint. Samples of a brown-red rust colored corrosion product and an underlying blue corrosion product taken from the underside of the foot were also analyzed by X-ray powder diffraction and were identified as cuprite and azurite respectively.

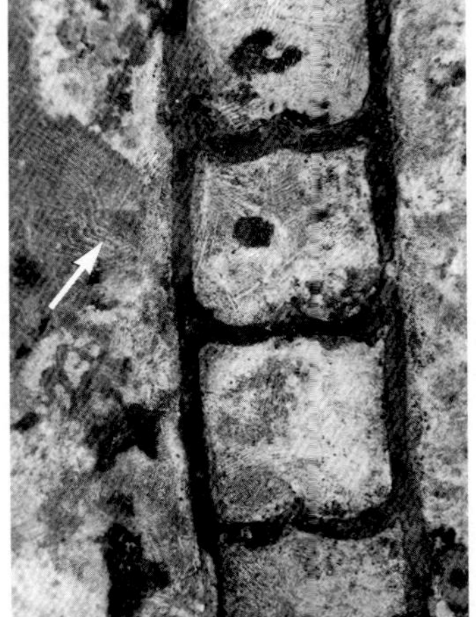

Fig. 8.2

Fig. 8.3

Figure 8.1. Interior base, magnified approximately 8x. Light circular rings are remnants of malachite bubbles broken during cleaning. Azurite particles that originally formed inside a malachite bubble are now exposed in the ring left of the center.

Figure 8.2. View of the bottom. The dragon is executed in intaglio. The magnified area (Fig. 8.3) is marked with an arrow.

Figure 8.3. View of the bottom. The dendritic structures are visible as light and dark striations in and around the dark intaglio scales of the dragon on the bottom on the *zhi*. The area is magnified approximately 8x.

Fanglei
No. 24 (2:1941)

The *fanglei* was cast in a four-piece mold assembly that separated at the corners. There were two separate molds, one for the lid and one for the body. There were probably eight divisions in the vessel mold: for example, each side would be made of two panels split vertically down the center that would be luted together to form one side of the vessel.[66] Mold marks are visible on the bottoms of some of the flanges, especially the corner ones near the foot.

Fig. 9.1

They are also visible on the vessel shoulder at the corners in the undecorated bands where the raised lines are misaligned. There are some on the edges of the rings on the side handles, especially the proper right one. Casting flash has been left on less accessible areas on the interior shoulder region of the vessel. The lid contains mold marks on the knob, stem, and flanges indicating that the mold split the lid in half lengthwise and at the four corners. There are two squarish depressions, one on each side on the underside of the knob (Fig. 9.1).[67] Their function was probably to allow for the escape of gases generated during casting.

Fig. 9.2

Many parts of the vessel have been integrally cast, including the masks, flanges, and the knob and stem on the lid. The three D-shaped handles with the antlered animal heads and solid handle rings were precast and set into the mold. The vessel was then cast onto them. Metal overflow is visible on the handles where they join the vessel (Fig. 9.2). The handles are open at the back and are filled with core material that is gray on the bottom handle and reddish on the two side handles. There are protuberances of extra metal on the vessel interior where the handles are located, to make the walls thicker to support them. This was accomplished during manufacture by hollowing out the inner core for the vessel to accept more metal during casting (Fig. 9.7).

There are several areas that seem problematic in casting the vessel, especially in the shoulder region. During casting the metal did not flow properly around the mask, resulting in a flaw. It was probably repaired during the time of manufacture by pouring additional metal in the area. A metal plug is present on the inside (Figs. 9.3–4).

Figure 9.1. Detail of the underside of the pyramid-shaped knob on the lid showing an oval-shaped core extension at center. A vertical mold mark is visible near the center.

Figure 9.2. Side view of the lower handle. Metal overflow at top and bottom of handle joints occurred when precast handles were set into the mold and the vessel was cast onto them.

Figure 9.3. View of the shoulder region. The area directly above the mask contains a patch repair (light rectangle, center).

Figure 9.4. CT slice through the shoulder region showing rough excess metal inside the vessel from the repair patch.

Fig. 9.3

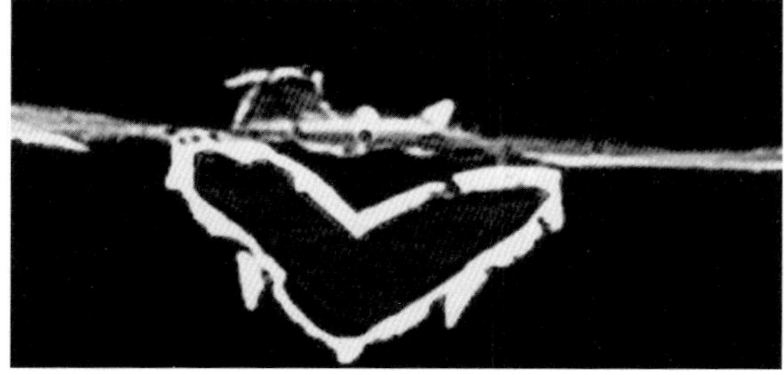

Fig. 9.4

Another fill occurs in the shoulder region on the opposite side, where there is a small metal hump, and there is a visible waver in the relief line decoration. On the vessel interior, small holes behind the masks on the shoulder may be associated with a process to get extra metal to flow into the mold when they were cast, or it could have been a means to allow for the venting of gas (Fig. 9.5).

Inscriptions are present on the inside of the lid and on the neck of the vessel. Each is made up of four incised lines: the two outer lines are continuous, and the two inner lines are divided into three segments. All appear to be integrally cast. One incised inscription line on the vessel was partly filled in during casting. There are two holes on the inscription on the lid. They occur where the metal is thinnest due to the incised exterior decoration and incised inscription. The irregular outlines around the holes are indicative of corrosion that probably occurred during burial, so the holes are most likely due to later corrosion rather than a casting flaw during manufacture.

Chaplets can be seen on the undecorated bands located directly above and below the shoulder on the vessel. Some are covered with iron corrosion products, possibly the result of ferruginous groundwater or materials associated with the object during burial. Chaplets are also present on the lid above the middle flanges.

The bronze's surfaces are covered with a dark green-black slightly glossy water patina mottled with overlying malachite and patches of matte granular blue-green copper corrosion products located mainly in crevices. Additionally the solid ring handles contain patches of cuprite with some whitish lead corrosion. Overall, the vessel contains intermittent encrustations of iron-rich pseudomorphs of bamboo matting, mainly on the bottom and relief areas on the sides, indicating the vessel was probably wrapped in a bamboo mat. The iron staining on the pseudomorphs may be due to groundwater, or elements in the soil, or they could be a lacquer-coloring agent that was present in the mat itself.[68]

Metal surfaces appear to have been lightly cleaned with acids, but not long enough to remove the malachite patina.[69] Surfaces fluoresce yellow under long-wave ultraviolet light due to a modern wax application. There is some pale orange fluorescence, most

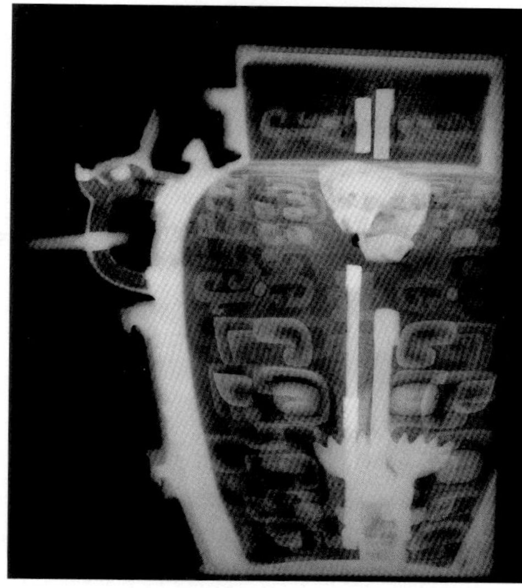

Fig. 9.5

Figure 9.5. X-radiograph of the body. Dark spots at the base of the masks indicate holes on the vessel interior that were vents and/or a means to get extra metal to flow into the mold. They are slightly offset from their actual location due to the angle of exposure during X-radiography.

Figure 9.6. DR of the *fanglei*. The knob on the lid, the animal handles, and the masks on the vessel are all hollow and contain cores.
(Elongation of the image is due to the pixel format of the CT imaging)

Fig. 9.6

notably on the ring handles, from lead corrosion products. The interior of both the vessel and lid are covered with dirt and velvety malachite and azurite copper corrosion products, as one would expect to find in closed containers that have been buried.[70]

The casting quality of the object is extraordinary in terms of size, amount of detail, and alignment of mold seams and workmanship. It is interesting to note that the corner flanges taper toward the center, more at the bottom than the top, though they look the same. Porosity in the casting is more concentrated at the neck, meaning that the object was cast bottom-up.

X-radiographs of the vessel bottom reveal four chaplets. Side views show the chaplets that can also be observed directly. The animal handles, masks, knob, and knob stem are hollow and contain core material. There are holes on interior walls associated with the masks, probably to allow gas to escape during casting (Fig. 9.5).[71]

Tomography of the vessel succinctly revealed the hollow core structure of the masks, handles, knob, and knob stem (Fig. 9.6). It showed that the cores in the pre-cast handles were hollowed out to accept additional metal, which appears as metal nubs on the handle interiors. An image produced from a model of the object shows this in a more three-dimensional view (Fig. 9.7). Another CT slice shows the casting flaw that was repaired by flowing in additional metal after the object was initially cast (Fig. 9.4).

Figure 9.7. Computerized model section through the chest region. The section cuts through the animal handles where they attach to the vessel below the circular rings. Both handles contain nubs of metal where the handle cores were hollowed out to accept additional metal prior to casting.

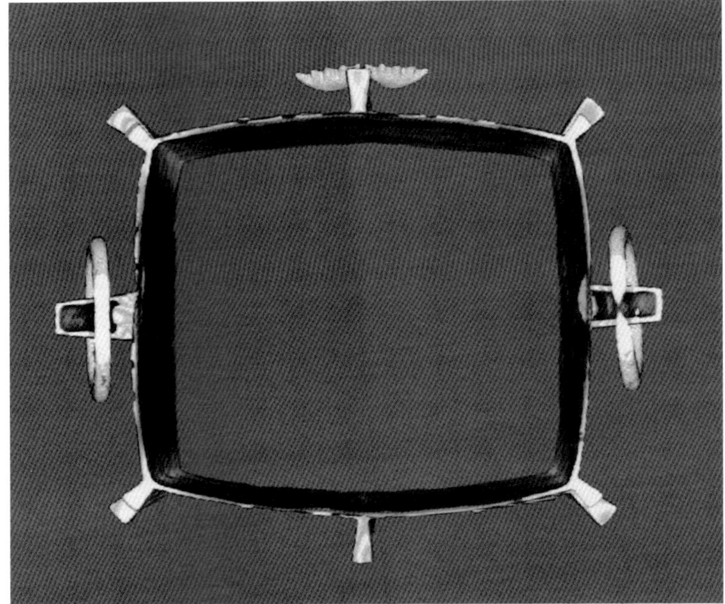

Fig. 9.7

Gui
No. 37 (223:1950)

Fig. 10.1

Figure 10.1. A view of the corner, showing a step and misalignment of the decorative wave pattern where the mold was sectioned. An oblong hole in the foot of the bowl (left of center) indicates that a chaplet has fallen out.

Figure 10.2. View of the underside. Ridges of casting flash run roughly through the center (A) and around the perimeter of the bowl (B). Note an oblong hole where a chaplet has fallen out (C).

Figure 10.3. Detail of the animal on the original handle. A squarish core extension is visible on the tail (A). A metal-filled hole is visible in the handle section near the bowl at right (B) where molten metal was poured to lock the handle in place.

Figure 10.4. CT slice through the foot of the bowl. Chaplets are indicated by dark spaces where they have been placed in a ring around the bowl. A gap shows where one chaplet has fallen out (arrow).

The *gui* was cast in a four-part mold that was sectioned diagonally at the corners. The square base and bowl were cast as one, with nubs protruding from the sides of the bowl for handle attachments. The handles were cast separately, then fastened to the bowl. Mold marks are evident on the corners and sides of the vessel where some of the design elements are misaligned and surfaces are out of plane (Fig. 10.1). The underside of the vessel is unfinished and rough with ridges of casting flash (Fig. 10.2). The inside bottom edge of the vessel has been left rough and thick on the interior to strengthen it. There are patches of reddish core material remaining on the underside of the vessel, in the square corners and in the recesses of the bowl. It contains gritty quartz and other particles as well as green and blue staining from copper corrosion. The original handle has a squarish 3 mm recess from a core extension or a vent opening on the tail of the small animal. Its purpose could be to register the core enclosed in the tail and to allow gas to escape (Fig. 10.3).

The handles were cast separately from the bowl. They were made in a two-part mold that was vertically bisected. There are mold marks running down the back neck of the dragon. The marks also appear as prominent ridges on the underside of the dragon's mouth and in the curled tail of the smaller animal attacking the dragon. There were separate mold inserts for the dragon's face and top of the head, identified by horizontal mold marks running across the brow and ears, and along the upper lip of the mouth. The handles are mainly hollow with pairs of holes near the ends for attachment to the vessel. They were attached by fitting them over exposed nubs on the bowl. Molten metal was poured into one hole, flowed around the nub, and exited the opposite hole, to lock the handle in place. Excess metal is present around the ends of the handles where they meet the vessel, and there is a protruding button of metal on one hole (Fig. 10.6).

Squarish and rectangular chaplets were carefully placed, mainly in undecorated areas on the sides of the vessel and in the bottom of the bowl. Eight chaplets were placed in a ring around the foot of the bowl; a hole is present where one has fallen out (Fig. 10.4).

The overall bronze surface is composed of a smooth gray-green water patina. There are intermittent encrustations of red, black, blue, and green granular copper corrosion products. The underside in particular is encrusted with red granular regions of cuprite and patches of malachite and azurite with some whitish corrosion from tin or lead. The inside of the bowl contains a thick compact layer of azurite that has the appearance of bright blue paint. Negative pseudomorphs of bird feathers are present where a bird or feather object lay in the bowl (Fig. 10.5). In general, the blue azurite follows a skewed hemispherical ring as if the corrosion formed while the vessel was tilted.

Fig. 10.2

Fig. 10.3

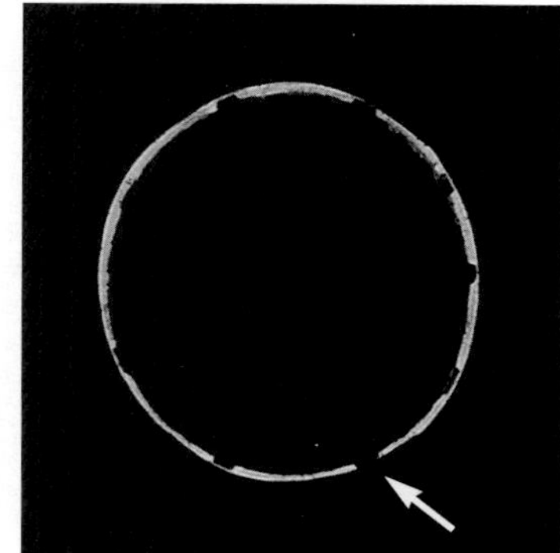

Fig. 10.4

Figure 10.5. Interior view of the bowl magnified roughly 5x. Remnants of feathers have been preserved in the azurite corrosion products.

Figure 10.6. Detail view of the original handle where it attaches to the vessel. Molten metal was poured in through holes in the sides of the handles. Excess metal is visible at one hole (A), and at the joint between the handle and vessel (B).

Figure 10.7. A CT slice through the *gui* where the handles are attached.

Figure 10.8. A detail of the CT slice through one handle. A dark line probably resulting from metal shrinkage clearly defines the nubs of metal that were cast as one with the bowl (A) and the surrounding metal (B) that was poured in through holes (C) to lock the handles in place.

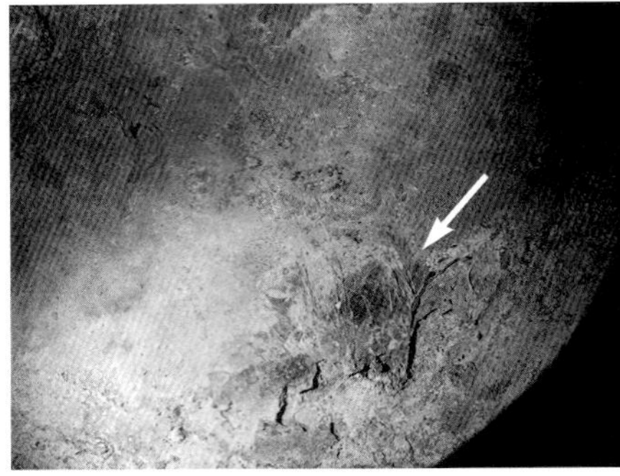

Fig. 10.5

The formation of the azurite is in keeping with observations of azurite formation on the interior of covered vessels.[72]

The vessel shows extensive breakage and repair with solder, adhesives, plaster fills, and overpaint. Repaired areas, especially the handles, fluoresce a bright yellow-green and dull to bright orange, which is characteristic of natural resins and shellac. X-radiographs indicate the side panels and portions of the bowl have been tacked together and repaired with solder. The handles are badly broken and have been repaired with glue and solder. One handle appears to be a more modern replacement. It is markedly different in design and execution, both visually and in the X-radiograph. Visually, the incised decoration is looser on the dragon's head and the incisions and metal edges on the small animal's tail are much more sharply defined. In the X-radiograph, the neck section containing the small animal has uniform density, indicating the metal is of recent origin. The bottom part of the handle is secured with a metal ring and the top is soldered to the dragon head. The dragon head is less dense in the X-radiograph, with little evidence of corrosion characteristic of ancient metal. The other handle, which appears to be original, is repaired with a spring or threaded rod inserted into the hollow interior for support. Both handles contain plaster fills and patches.

Other features revealed in the X-radiographs are chaplets in the sides of the square base and the bottom of the bowl, and a ring of chaplets surrounding the bowl. The extent of corrosion is also visible in the varying density of the X-radiograph where the tin component of the bronze is heavily corroded and appears dark. This is indicative of Type I corrosion where the high-tin delta phase of the metal corrodes first, leaving the copper-rich alpha phase behind. The dendritic structure of the metal is visible.

Tomography done on the sides of the bowl clearly revealed details about the handle attachment. A CT slice through the handle at its point of attachment to the bowl shows how the molten metal flowed around the nub of the vessel to secure the handle in position (Figs. 10.6–8).

Fig. 10.6

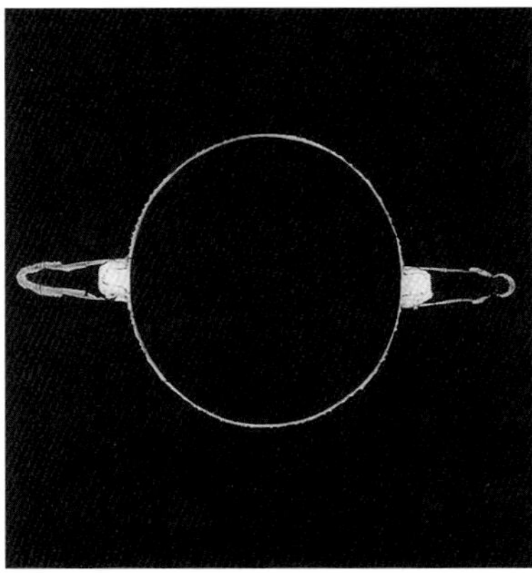

Fig. 10.7

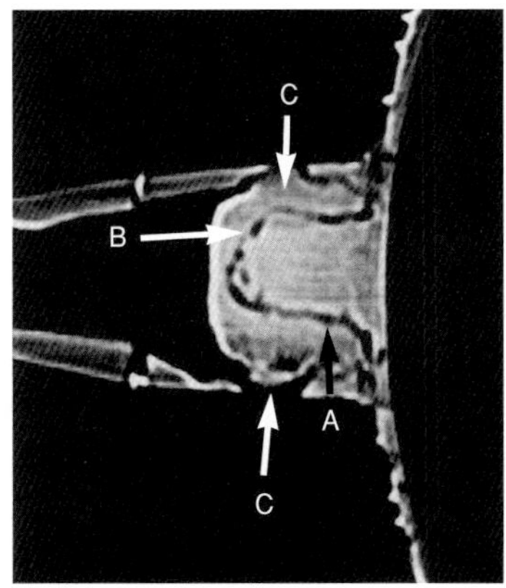

Fig. 10.8

Acknowledgments

I am indebted to W. Thomas Chase, Head Conservator at the Freer Gallery of Art and Arthur M. Sackler Gallery for his insights and wealth of knowledge and experience in all aspects of Chinese bronze conservation and technology. I am especially appreciative of the time he spent examining our bronzes at Saint Louis and the Freer, and his instruction in X-radiography and XRF work. I wish to express my sincere gratitude to my colleague Diane Burke for her assistance with X-radiography of the bronzes as well as her encouragement and support during all phases of this project. Instrumental to this work were conservation, analytical, and curatorial staff at the Freer and Sackler galleries: Paul Jett, Supervisory Conservator, for his examination and metal sampling; Janet Douglas, Conservation Scientist, for her XRD work and X-radiography instruction; and Jenny So, Curator, for her thought-provoking curatorial insights.

My warm thanks also to Ben Nightingale, X-Ray Applications at GE Aircraft Engines, with assistance from Joe Portaz, and Warren Sipos for Industrial X-ray Computed Tomography; Bob Vocke, Research Scientist at NIST for lead isotope analysis; Dan Kremser, Microprobe Specialist at Washington University for SEM analysis; and Zhou Bao Zhong, Director of Conservation at the National Museum of Chinese History, for his thoughtful examination of the bronzes in St. Louis. My appreciation to conservators in private practice Cathy Coho, for her examination and information on textile pseudomorphs, and Chris Augerson, for his assistance with IR spectroscopy; and to Zoe Perkins, Textile Conservator, The Saint Louis Art Museum, for her examination of fiber pseudomorphs.

I am grateful for the opportunity to examine these magnificent bronzes. It has been a privilege to treat them in the ongoing study of their creation and to help preserve them for future generations to appreciate.

Notes

All photographs in this section, with the exception of tomography images, are by Suzanne Hargrove.

1. See Chase 1991b, 86.

2. There is no evidence to suggest that metalworking processes such as forging or sheet metal work were used to create vessels prior to the middle to late Zhou dynasty. This may be due in part to the alloy compositions of the bronzes, which could not be worked because they were too brittle. See Chase 1983, 104, 110–11; Meyers and Holmes 1983, 124; Scott 1991, 26.

3. There is some question as to whether a model was necessary to create designs in the mold. Most scholars believe the decoration in the earliest Shang bronzes was done by drawing directly in the mold without using a model (Loehr Style I and II). Many feel the ensuing styles (Styles III–V) necessitated the use of a model and that designs were drawn on both the mold and the model. Lastly, in the middle and late Zhou and Han dynasties other techniques were introduced including pattern-stamping of molds, lost-wax casting, multiple-mold casting, and the creation of permanent metal molds that could be reused. See Bagley 1980, 70–73, 182; Bagley 1987, 37–41; Barnard and Satō 1975, 3–6; Chase 1983, 112, 116–17; Chase 1991a, 32; Freestone, et al. 1989, 270; Meyers and Holmes 1983, 124–25, 129–31.

4. Loess is a local windblown rock dust of fine yellow earth. By itself, loess is friable without much cohesion. When used as a modelling material it could have been mixed with clay or an organic binder to make it more plastic and cohesive. For molds, it has been suggested that a liquid organic binder, which would leave no fossil evidence, was added to impart strength prior to firing. Clay would not have been added as it would change the properties that make loess a desirable mold material: its fineness, which made it both easy to carve and able to capture minute detail for mold impressions; its dimensional stability during drying and firing; and its high porosity that enabled it to absorb gases generated during casting. See Chase 1983, 113; and Freestone, et al. 1989, 257–58.

5. Chase 1983, 113.

6. Freestone, et al. 1989, 258.

7. Chase 1983, 115. It is uncertain whether the molds were fired prior to casting the molten metal, or whether they were vitrified by the pour (Freestone, et al. 1989, 270).

8. Chase 1991b, 91.

9. Ibid.

10. Chase 1983, 115.

11. Chase 1991b, 91; and Freestone, et al. 1989, 258.

12. For a detailed discussion of cores see, among others, Bagley 1987, 41.

13. Meyers and Holmes 1983, 125.

14. Gettens 1969, 69–70; and Meyers and Holmes 1983, 133–34.

15. For copper cores, the solid copper was precast and probably held in the mold via extensions on the copper that were cut off after casting. It is thought the copper cores prevented excess metal shrinkage in the legs upon cooling. See Chase 1991a, 32; and Meyers and Holmes 1983, 134.

16. Some are scrap from other bronze vessels. Chaplets can sometimes be detected on the bronzes or in X-radiographs of the bronzes as shown in this catalogue (Gettens 1969, 98–107).

17. Meyers and Holmes 1983, 132.

18. Some evidence indicates the molds were not preheated. Examination of the metal microstructure of the cast bronze surrounding chaplets showed a chill zone where the molten metal solidified against a cold chaplet, meaning the molten metal was poured into molds that were not preheated (Gettens 1969, 132–34). In other instances the molds might have been heated where metal crystal formations show that "chill zones are usually not apparent, confirming the lack of a chilling effect of the ceramic mold" (Chase 1991b, 89–90).

19. Chase 1991a, 23. An in-depth discussion of Chinese bronze-casting foundry methods can be found in Barnard and Satō 1975.

20. Chase 1991b, 91.

21. A more precise term would be "mold flash" or "casting flash," but "mold marks" is the most common reference.

22. Chase 1991b, 93.

23. Bagley 1987, 41; and Chase 1991b, 94.

24. Chase 1983, 102.

25. Gettens 1969, 130. To gain a better understanding of metal corrosion, it is helpful to have a basic understanding of the structure of metals. An excellent resource on the subject is David A. Scott's *Metallography and Microstructure of Ancient and Historic Metals* (Scott 1991), which has been of immeasurable assistance in this publication.

26. Chase 1991b, 97; Gettens 1969, 177.

27. Chase 1991b, 99–103. Recent personal communication from W.T. Chase, Head of the Freer Gallery of Art/Arthur M. Sackler Gallery Department of Conservation and Scientific Research, Washington, D.C., is that alpha removal and delta removal are more precise and less confusing.

28. These layers can be sequential, consisting of an initial layer of red cuprite possibly mixed with whitish tin oxides (called a cuprite marker layer) that forms on and can define the original surface of the bronze. This layer is followed by the emerald-green malachite and blue azurite copper corrosion products as well as whitish tin and whitish lead corrosion products. In the metal microstructure, the tin-rich phase of the eutectic is the first area to corrode, followed by the remaining alpha phase. See Chase 1991b, 102; and Gettens 1969, 129–31, 172–86.

29. This has been termed an "anionic control" process in which anodic ions (oxygen and chloride) migrate into the bronze structure (Chase 1991b, 101).

30. Chase 1991 b., 103.

31. The water patina can be loose and friable, or powdery white, but usually it is semiglossy and forms a hard shell-like coating (1 mm or more) over the surface like lacquer that can chip off. In the metal microstructure the alpha phase of the alloy dissolves while the delta remains (Chase 1991b, 103; and Gettens 1969, 186–88).

32. Formation of a cohesive tin-oxide patina is thought to be affected by a "cationic control" process in which cations (copper I and II) slowly migrate to the surface of the bronze (Chase 1991b, 101).

33. It is thought that the water patina is the product of a protective patina that was intentionally applied to the bronze in antiquity and subsequently determined the type, progress, and extent of corrosion that occurred during burial. This is borne out by bronzes with evidence of Type I corrosion occurring along areas of freshly exposed bronze where they sustained damage in the form of cracks, breaks, or pits during burial (ibid., 103).

34. The earliest publications describing the use of industrial computed tomography for the study of art and archaeological material date from the 1980s. Examples include: Avril and Bonadies 1991; Cheng and Mishara 1988; Conroy and Vannier 1984; Tout, et al. 1980; Unger and Perleberg 1987.

35. Chase 1983, 116.

36. Gettens 1969, 186.

37. In antiquity the intaglio design would have been created by drawing or stamping into a plastic material, resulting in a raised edge (Gettens 1969, 141–43).

38. Chase 1991a, 26.

39. Ibid.

40. The carbon could be remnants from the vessel's use, where the *gu* would have been placed over a fire to heat its wine contents.

41. The delta phase has mineralized to black in the gray portions of patina, and to a lighter turquoise-green in the turquoise areas of patina. The color differences may be the result of different states of tin oxidation (observation by W.T. Chase upon examination of the bronze).

42. Chase 1991b, 103.

43. Chase 1991a, 26.

44. There is still speculation as to how inscriptions were done (Chase 1983, 115).

45. Gettens 1969, 118.

46. See FTIR data on black material removed from *fangding* 222:1950 (No.16). Further research is planned to study this coating.

47. Observation by Cathy Coho, textile conservator in private practice.

48. Gettens 1969, 114.

49. Their function may have been to register the cores inside the legs, and/or to allow gas to escape when the metal was poured (Gettens 1969, 113).

50. Black material also present around the inscription is probably a product left from rubbings.

51. SEM (scanning microscopy analysis) using a JEOL electron microscope courtesy of Dan Kremser, microprobe specialist at Washington University, St. Louis; FTIR (Fourier Transform Infra red Spectra) analysis courtesy of Chris Augerson, St. Louis.

52. SEM will not detect carbon which may be present in a lacquer inlay and/or from the vessel's use over a fire in ancient times (X-rays from carbon are absorbed by the thick detector window). SEM analysis confirmed the presence of copper, tin, lead, and chlorine as the major constituents with low amounts of phosphorous and some sulfur. There were trace amounts of zinc, nickel, and iron.

Additionally, one sample contained silica, alumina, magnesia, and sodium. The presence of silica is promising as an intentional inlay ingredient.

53. The indefinite results may be due to components added during lacquer processing, breakdown of the material over time, or environmental factors.

54. Fibers and pseudomorphs were examined by Cathy Coho, textile conservator in private practice.

55. Zoe Perkins, Head of the Textile Conservation Department, The Saint Louis Art Museum.

56. Observation by Cathy Coho.

57. Meyers and Holmes 1983, 125–27. Further examination indicates that the effect does not appear to be mold splash, since it occurs on more than one side of the bronze.

58. Analysis performed by Janet Douglas at the Freer/Sackler Gallery Department of Conservation and Scientific Research, using a Philips XRG-2600 X-ray powder diffraction unit.

59. Chase 1991b, 103.

60. Chase 1991a, 33; and Gettens 1969, 70–74.

61. This material has yet to be analyzed.

62. The copper cores would help prevent excess shrinkage in the legs, which could otherwise lead to casting failure (Chase 1991a, 33).

63. The dendrites appear light, as they have been converted to tin oxide (Gettens 1969, 126–27).

64. Analysis performed in September 1971 by W.T. Chase and I.V. Bene. Results indicated a metal composition of 76% copper, 16% tin, and 5% lead. This would be considered an appropriate alloy composition for this time period.

65. Analysis performed by Janet Douglas at the Freer/Sackler Department of Conservation and Scientific Research. Samples were mounted with rubber cement on glass fibers. They were analyzed using a Phillips XRG-2600 X-ray powder diffraction unit.

66. Observation by W.T. Chase.

67. They are reminiscent of the core extensions found on *fangding* No. 15 (108:1954) and No. 16 (222:1950). Perhaps their function is to register the core and/or allow gas to escape.

68. Observations based upon examination by W.T. Chase and Jenny So at the Freer Gallery of Art/Arthur M. Sackler Gallery.

69. If a mechanical cleaning method had been used, the malachite bubbles comprising the patina would have been broken.

70. Gettens 1969, 178.

71. The bronzes studied reveal that those with features containing cores have core extensions or holes, leading me to believe that appendages with cores must have vent holes to allow gas to escape and register the core.

72. Gettens 1969, 178.

REFERENCES

Araki and Satō 1978 — Araki Tadashi and Satō Hisayoshi. 1978. "Exhibition Lighting and Lacquered Ware Discoloration" [Japanese text with English summary]. *Kobunkazai no kagaku* 23:1–24.

Avril and Bonadies 1991 — Avril, Ellen B., and Stephen D. Bonadies. 1991. "Non-Destructive Analysis of Ancient Chinese Bronzes Utilizing Industrial Computed Tomography." In *Materials Issues in Art and Archaeology II*, ed. Pamela B. Vandiver, et al., 49–63. Materials Research Society Symposium Proceedings, vol. 185. Pittsburgh: Materials Research Society.

Badler, et al. 1990 — Badler, Virginia R., Patrick E. McGovern, and Rudolph H. Michel. 1990. "Drink and be Merry!: Infrared Spectroscopy and Ancient Near Eastern Wine." In *Organic Contents of Ancient Vessels: Materials Analysis and Archaeological Investigation*, ed. William R. Biers and Patrick E. McGovern, 25–36. MASCA Research Papers in Science and Archaeology, vol. 7. Philadelphia: University of Pennsylvania.

Bagley 1980 — Bagley, Robert W. 1980. "The Beginnings of the Bronze Age: The Erlitou Culture Period and the High Yinxu Phase (Anyang Period)." In *The Great Bronze Age of China: An Exhibition from the People's Republic of China*, ed. Wen Fong, 67–96, 182. New York: The Metropolitan Museum of Art.

Bagley 1987 — ————. 1987. *Shang Ritual Bronzes in the Arthur M. Sackler Collections*. Ancient Chinese Bronzes in the Arthur M. Sackler Collections, vol. 1. Cambridge: Harvard University Press.

Barnard and Satō 1975 — Barnard, Noel, and Satō Tamotsu. 1975. *Metallurgical Remains of Ancient China/Chūgoku kodai kinzoku ibutsu*. Tokyo: Nichiōsha.

Bonadies 1991 — Bonadies, Stephen D. 1991. "Tomography of Ancient Bronzes." In *Ancient and Historic Metals: Conservation and Scientific Research*, ed. David A. Scott, et al., 75–84. Marina del Rey: The Getty Conservation Institute.

Chase 1983 — Chase, W. Thomas. 1983. "Bronze Casting in China: A Short Technical History." In *The Great Bronze Age of China: A Symposium*, ed. George Kuwayama, 100–23. Los Angeles: Los Angeles County Museum of Art.

Chase 1991a — ————. 1991. "Ancient Chinese Bronze Art: Casting the Precious Sacral Vessel." In *Ancient Chinese Bronze Art: Casting the Precious Sacral Vessel*, 19–38. New York: China House Gallery, China Institute of America.

Chase 1991b — ————. 1991. "Chinese Bronzes: Casting, Finishing, Patination, and Corrosion." In *Ancient and Historic Metals: Conservation and Scientific Research*, ed. David A. Scott, et al., 85–117. Marina del Rey: The Getty Conservation Institute.

Cheng and Mishara 1988 — Cheng, Yu-Tarng, and Joan Mishara. 1988. "A Computerized Axial Tomographic Study of Museum Objects." In *Materials Issues in Art and Archaeology*, ed. Edward V. Sayre, et al., 19–24. Materials Research Society Symposium Proceedings, vol. 123. Pittsburgh: Materials Research Society.

Conroy and Vannier 1984 — Conroy, Glenn C., and Michael W. Vannier. 1984. "Non-Invasive Three-Dimensional Computer Imaging of Matrix-Filled Fossil Skulls by High-Resolution Computer Tomography." *Science* 226(4673):456–58.

Dennis 1989 — Dennis, Michael J. 1989. "General Electric Company, NDE Systems and Services, Industrial Computed Tomography." In *Nondestructive Evaluation and Quality Control*, 358–86. The Metals Handbook, vol. 17. 9th ed. Materials Park, Ohio: ASM International.

Derrick, et al. 1988 — Derrick, Michelle, C. Druzik, and Frank Prusser. 1988. "FTIR Analysis of Authentic and Simulated Black Lacquer Finishes on 18th-Century Furniture." In *Urushi: Proceedings of the Urushi Study Group, June 10–27, 1985, Tokyo*, ed. N.S. Bromelle and Perry Smith, 227–34. Marina del Rey: The Getty Conservation Institute.

Du 1988 — Du Yamin. 1988. "The Production and Use of Chinese Raw Urushi and the Present State of Research." In *Urushi: Proceedings of the Urushi Study Group, June 10–27, 1985, Tokyo*, ed. N.S. Bromelle and Perry Smith, 189–95. Marina del Rey: The Getty Conservation Institute.

Freestone, et al. 1989 — Freestone, Ian C., N. Wood, and Jessica Rawson. 1989. "Shang Dynasty Casting Molds from North

China." In *Cross-Craft and Cross Cultural Interactions in Ceramics*, ed. Patrick E. McGovern, 253–73. Westerville, Ohio: American Ceramics Society, p. 270.

Gettens 1969 — Gettens, Rutherford J. 1969. *The Freer Chinese Bronzes, Vol. II: Technical Studies*. Washington, D.C.: Smithsonian Institution.

Gianno 1990 — Gianno, Rosemary. 1990. "Archaeological Resins from Shipwrecks Off the Coasts of Spain and Thailand." In *Organic Contents of Ancient Vessels: Materials Analysis and Archaeological Investigation*, ed. William R. Biers and Patrick E. McGovern, 59–67. MASCA Research Papers in Science and Archaeology, vol. 7. Philadelphia: University of Pennsylvania.

Hanson and Pell-Walpole 1951 — Hanson, D., and W.T. Pell-Walpole. 1951. *Chill-Cast Tin Bronzes*. London: Edward Arnold and Co.

Institute of the History of Natural Sciences 1983 — Institute of the History of Natural Sciences, Chinese Academy of Sciences. 1983. *Ancient China's Technology and Science*. Beijing: Foreign Languages Press.

Karlbeck 1935 — Karlbeck, Orvar. 1935. "Anyang Moulds." *Bulletin of the Museum of Far Eastern Antiquities, Stockholm* (7):39–60.

Kenjo 1977 — Kenjo Toshiko. 1977. "Effect of Humidity on the Hardening of Lacquer." *Proceedings, First International Symposium on the Conservation and Restoration of Cultural Property: Conservation of Wood*, 151–63. Tokyo: Tokyo National Research Institute of Cultural Properties.

Kenjo 1978 — ————. 1978. "Infrared Spectrometry of Lacquer Films" [Japanese text with English summary]. *Scientific Papers on Japanese Antiques and Arts Crafts* 23:32–39.

Kumanotani 1983 — Kumanotani Ju. 1983. "Japanese Lacquer: A Super Durable Coating." In *Polymer Application of Renewable Resource Materials*, ed. Charles E. Carraher and L.H. Sperling, 225–48. Polymer Science and Technology, vol. 17. New York and London: Plenum Press.

Kumanotani 1988 — ————. 1988. "The Chemistry of Oriental Lacquer (*Rhus Verniciflua*)." In *Urushi: Proceedings of the Urushi Study Group, June 10–27, 1985, Tokyo*, ed. N.S. Bromelle and Perry Smith, 243–51. Marina del Rey: The Getty Conservation Institute.

Kumanotani, et al. 1978 — Kumanotani Ju, Achiwa Munro, Oshima Ryuivhi, and Adachi Kimikazu. 1978. "Attempts to Understand Japanese Lacquer: Japanese Lacquer as a Super-Durable Material." *Proceedings: International Symposium on the Conservation and Restoration of Cultural Property, Cultural Property and Analytical Chemistry*, 50–62. Tokyo: Tokyo National Research Institute of Cultural Properties.

Meyers and Holmes 1983 — Meyers, Pieter and Lore L. Holmes. 1983. "Technical Studies of Ancient Chinese Bronzes: Some Observations." In *The Great Bronze Age of China: A Symposium*, ed. George Kuwayama, 124–36. Los Angeles: Los Angeles County Museum of Art.

Scott 1991 — Scott, David A. 1991. *Metallography and Microstructure of Ancient and Historic Metals*. Marina del Rey: The Getty Conservation Institute and Santa Monica: The J. Paul Getty Museum.

Tout, et al. 1980 — Tout, R.E., W.B. Gilboy, and A.J. Clark. 1980. "The Use of Computerized X-Ray Tomography for the Non-Destructive Examination of Archaeological Objects." In *Proceedings of the 18th International Symposium on Archaeometry and Archaeological Prospection*, 608–16. Cologne: Rheinland-Verlag.

Tylecote 1983 — Tylecote, R.F. 1983. "Ancient Metallurgy in China." *The Metallurgist and Materials Technologist* 15(9):435–39.

Unger and Perleberg 1987 — Unger, Achim, and Jens Perleberg. 1987. "X-Ray Computer Tomography (XCT) in Wood Conservation." In *ICOM Committee for Conservation: 8th Triennial Meeting*, ed. Kristen Grimstad, 99–105. Marina del Rey: The Getty Conservation Institute.

Glossary of Technical and Conservation Terms

Alloy — A mixture of two or more metals. Alloying refers to the intentional addition to a metal of other metal(s) to obtain certain qualities. Binary alloys are a mixture of two metals. Ternary alloys are a mixture of three metals; a Chinese *bronze* could contain copper, tin, and lead.

Alpha phase — A copper-rich solid solution of tin in copper.

Atacamite — Basic copper chloride: $CuCl_2 \cdot 3Cu(OH)_2$. A copper corrosion product that occurs as pits, crusty patches, or blisters on *bronze*. It often is pale green in color with a powdery or granular consistency, and can be found in combination with other copper corrosion products such as *malachite*, *cuprite*, or *azurite*. It is associated with an ongoing destructive corrosion process called "bronze disease" that is active under humid conditions.

Azurite — Blue carbonate of copper: $2CuCO_3 \cdot Cu(OH)_2$. It is thought to form under drier conditions than *malachite*, hence it is usually seen on the interior of covered vessels. It can be thin and compact having the appearance of blue paint, or can have a more granular velvety crystalline appearance.

Beta phase — A phase that develops during cooling of molten *bronze* below 800° C in bronzes containing more than 8% tin. The beta phase is the product of a reaction between the *alpha phase* and remaining tin-rich liquid in the melt.

Bronze — An *alloy* of copper and tin. A low-tin bronze contains less than 17% tin. Chinese bronzes were additionally alloyed with lead to varying degrees.

Casting — A process in which molten metal is poured into a prepared mold and allowed to solidify.

Cassiterite — Stannic oxide: SnO_2. On high-tin *bronzes* it often appears as a minty green lacquerlike *water patina*, though at times it can be loose and friable or powdery white.

Cerussite — Lead carbonate: $PbCO_3$. Cerussite appears as a grayish crust on heavily corroded *bronzes*. It can be dense and white so that it can be mistaken for tin oxide; it can also be reddish in color when mixed with litharge (lead monoxide). It can be found in combination with *malachite*, or can form a distinct layer beneath it. It fluoresces pink under long-wave ultraviolet light. Like *cuprite*, it can fill voids left by corroded lead globs in the bronze *alloy*; looking like round white pockets in metal cross sections.

Chaplets — Metal spacers that are placed in a mold to separate it from the inner *core*. The chaplets maintain the void for the molten *bronze* to fill during *casting* and are incorporated into the molten bronze.

Core — A solid form, usually made of fine-grained earth called loess, that occupies the interior portions of a *bronze* during *casting*. It defines the interior surface walls of the bronze.

Core extension — *Core* material that extends from the core outward to the mold, or to another core such as an *interleg core*. Its function may be to register the position of the core and allow for the escape of gas generated during *casting*. Core extensions are visible as rectangles on the insides of legs in *ding*, and where crosses are present in *gu*.

Coring — A condition that occurs in *dendrite* crystal formations in the microstructure of a metal. As a molten metal *alloy* cools and begins to solidify, a compositional gradient occurs between the interior core and the exterior "skin" of the dendrite. In a *bronze*, the inner core of the dendrite will be richer in copper since it has the higher melting point and solidifies first (melts at 1083° C). The outer surface of the dendrite will be richer in tin, which has the lower melting point (melts at 232° C).

Cuprite — Copper oxide: Cu_2O. A copper corrosion product that ranges from bright orangish to dark red in color. It can be compact and matte, or have a granular appearance like tiny garnet crystals on a *bronze*. It is often found under green basic copper carbonate corrosion products, where it lies adjacent to uncorroded metal (see *cuprite marker layer*). It can have the appearance of bright orange-red paint inside incised decoration, where it has been mistaken for inlay. In metal cross sections it can appear as round inclusions where it has replaced voids left by lead globules that have corroded away in the bronze.

Cuprite marker layer — A tarnish film of copper I oxide (Cu_2O) that forms on the original surface of an antiquity by oxidation of the copper component of an *alloy* upon exposure to moist air. It can serve to identify the location of an original metal surface on a heavily corroded object.

Delta phase	An intermetallic compound of fixed composition: $Cu_{31}Sn_8$. It occurs in *bronzes* containing 5–15% tin (or more, depending on cooling conditions). It is light blue, hard, and brittle, often with a jagged appearance.
Dendrite	A type of crystal growth in a metal *alloy* that resembles a fern or tree. Sometimes they can be seen in the metal with an unaided eye or with low magnification, otherwise they may only be visible through polished and stained metal cross sections. Dendrites are common in Chinese *bronzes*.
Eutectic	A metal microstructure of fixed composition resulting from the partial solubility of two metals in one another. It is the *alloy* composition with the lowest melting point. Upon cooling the liquid metal goes directly to a solid. For binary alloys the eutectic often consists of fine plates of *alpha* and *beta phases* evenly dispersed in one another.
Eutectoid	A metal microstructure produced when one solid *phase* decomposes into two finely dispersed solid phases upon cooling. In tin *bronzes* the eutectoid is composed of the *alpha* and *delta* phases. The eutectoid appears in the microstructure of 5–15% Sn and up. It is a bright blue, hard, and brittle metal that has a jagged appearance. The structure is shaped by grain boundary edges.
Gamma phase	A transitional *phase* occurring in the formation of a *eutectoid* metal structure. As a molten *bronze* cools, the *beta phase* transforms to gamma, which upon further cooling transforms to a final solid mixture of *alpha* plus *delta* eutectoid. It occurs in bronzes containing 5–15% tin or more, depending on cooling conditions.
Hot tearing	Coarse cracks that occur in shaped *castings* as a result of local restriction of the metal during cooling.
Intaglio	Incised decorations that are below the metal surface of the *bronze*.
Interleg core	An interior *core* section that defines the inside of the legs, and bottom of a vessel.
Integral casting	Components of a vessel cast as one with the vessel in a single pour of molten metal. This may include appendages such as knobs, stems, flanges, and masks, as well as legs and handles.
Karlbeck lines	Geometric or pictorial relief lines that occur on the bottoms of Chinese *bronzes*. Their possible function may have been to allow for the escape of gas generated during *casting* and/or to register the alignment of the foot *core*; they may have also defined the surface of the core before it was shaved down to create the casting space for the vessel. They are named after the Swedish engineer and archaeologist Orvar Karlbeck.
Loess	A windblown rock dust of fine yellow earth that is characteristic of the North China Plain. It was used to create molds, models, and *cores* for Chinese *bronzes*.
Malachite	Green carbonate of copper: $CuCO_3 \cdot Cu(OH)_2$. A green copper corrosion product that occurs on ancient *bronzes*. It is usually found in crusty patches over *cuprite*, but can also appear as a hard continuous layer, or as bright green layer like enamel paint.
Mold marks	More appropriately called "mold flash" or "*casting* flash"; marks made where molten metal ran into joints between mold sections. On cast metal objects they appear mostly as vertical narrow ridges of metal. Often they were removed in finishing operations and so can only be found in areas that were difficult to reach, or in recessed areas that were hidden from view.
Patina	The surface appearance of a *bronze* produced either intentionally by an artist/craftsman, or as a result of corrosion that has occurred over time. A Chinese bronze may contain evidence of both.
Phase	One component of a metallic system that is homogenous and of uniform composition throughout. A solid crystal form. Phases are often designated by Greek letters; those frequently discussed for Chinese *bronzes* include *alpha*, *beta*, *gamma*, and *delta*, which have different properties. Phases can be mutable under changing temperatures as molten bronze cools. Where more than one phase is present, a bronze *alloy* can show partial solubility of metals in one another which can produce different microstructures including *eutectic* and *eutectoid* structures commonly found in Chinese bronzes.
Piece-mold	A *casting* method that uses a mold assembled from sections that are fitted together. Mold fragments from Chinese *bronzes* often contain pyramid-shaped mortise and tenon joints.

Pseudomorph	Mineralized formations of organic material that are preserved in metal corrosion products, resulting from reactions between the metal, organic material, and components in the burial environment. Pseudomorphs may be positive (the result of the replacement of organic materials by metal corrosion products) or negative (impressions of decaying organic material left in metal corrosion products).
Seam lines	See *Mold marks*.
Segregation	Metal microstructures that form where components of an *alloy* separate out upon cooling. In alloys, three types of segregation are common: normal, dendritic, and inverse. In normal segregation, the lower-melting-point metal is concentrated toward the inner region of the *casting*. In dendritic, fernlike growth (*dendrites*) occurs from local compositional gradients. In inverse, lower-melting-point metal is concentrated toward the outer regions of a casting.
Septum	A dividing wall in a vessel, which separates the body from the foot and serves as the bottom of the vessel; typically found in *gu*.
Sprue	A broken-off ridge of metal left from a repair where molten metal has been used to fill a hole.
Water patina	A smooth, somewhat glossy, minty green *patina* found on Chinese *bronzes* of higher tin content. It is a thin layer of hydrous tin oxide resulting from removal of the copper-rich *alpha* phase of the bronze *alloy* and its replacement by tin oxide corrosion products without a change in volume. The greenish color is due to staining caused by residues of copper salts. Other copper corrosion products such as *malachite* and *azurite* can form on top of the patina where they can be mechanically removed. *Cuprite* is generally absent.

LEAD ISOTOPE RATIOS

No.	Accession Number	Vessel Type	NIST Number	Lead Isotope Ratio			Sampling Locations
				208/206	207/206	204/206	
1	29:1951	*gu*	2800	2.0178	0.77070	0.04847	bottom underside, two locations
5	227:1950	*lei*	2804	2.1107	0.85620	0.05470	Lid, solder removed from bottom edge
				2.1117	0.85634	0.05469	
				Average: 2.1112*	Average: 0.85627*	Average: 0.05469*	
8	217:1950	*gu*	2795	1.9417	0.74304	0.04639	Foot, bottom edge
14	125:1951	*zun*	2796	2.1479	0.87277	0.05604	Foot, bottom edge
15	108:1954	*fangding*	2799	2.0866	0.84707	0.05420	Flange, bottom
16	222:1950	*fangding*	2791	2.2042	0.89637	0.05796	Foot, underside
19	127:1951	*fangyi*	2797	2.1697	0.89036	0.05722	Lid, bottom edge
			2802	2.1695	0.8905	**	Vessel, bottom edge
20	221:1950	*jia*	2794	2.1678	0.89047	0.05736	Foot, underside
21	30:1985	*zhi*	2798	2.1241	0.85962	0.05513	Foot, underside
24	2:1941	*fanglei*	2793	2.1625	0.88777	0.05717	Lid, underside bottom
			2801	2.1685	0.89088	0.05738	Handle (proper right), underside
			2803	2.1689	0.89149	0.05747	Foot, underside
37	223:1950	*gui*	2792	2.1007	0.84960	0.05429	Bottom edge
				2.1013	0.84979	0.05427	
				Average: 2.1010*	Average: 0.84970*	Average: 0.05428*	

* sample tested twice to show range of variability; vessels selected at random

** very little lead, no ratio

The element lead exists as four isotopes: ^{204}Pb, ^{206}Pb, ^{207}Pb, and ^{208}Pb. Of these, ^{204}Pb is unstable and was formed primordially. The isotopes ^{206}Pb, ^{207}Pb, and ^{208}Pb are stable end products from the decay of uranium and thorium. Expressed as ratios, lead isotope determinations from separate geological deposits vary depending upon the method of formation and age.

Lead isotope ratios have great potential in the study of Chinese bronzes. Ratios obtained from ancient ore sites can be compared to those gathered from lead-containing artifacts, including Chinese bronzes, to identify probable geographical mining origins. This information can lead to further determinations of provenance, trade routes, and other socio-economic inferences. In addition, it has been discovered that lead isotope data among Chinese ritual bronzes can be used to categorize them by chronology and style, based on their parent lead sources.

Eleven bronzes were sampled as part of this technical study. In some cases two or more samples were taken from the same object (as with lidded vessels), resulting in a total of fourteen samples. Roughly 15 mg of bronze were obtained for each sample using a tungsten carbide drill. The surface corrosion layers from the drilling were discarded to avoid contamination of the sample. The samples were prepared by dissolving them in acid, and separating out the lead by anodic electrodeposition. The extracted lead was measured by mass spectrometers to obtain proportions of stable lead isotopes present which are expressed as ratios ^{208}Pb/^{206}Pb, ^{207}Pb/^{206}Pb, and ^{204}Pb/^{206}Pb. The procedure was calibrated against National Institute of Standards and Technology SRM #981. The values reported have an estimated uncertainty less than or equal to 0.05% relative.

The samples were taken by Paul Jett, Objects Conservator at the Freer Gallery of Art. The separations and analysis were performed by Robert Vocke, Research Chemist at the National Institute of Standards and Technology (formerly the National Bureau of Standards). The data obtained in this study will add to lead isotope databases of Chinese bronzes from this and other study collections. The assembled data, combined with analytical, archaeological, and art historical assessment will shed new light on provenance and stylistic, socio-economic, and technological developments associated with Chinese bronze production.

References

Barnes, I. Lynus, W. Thomas Chase, and Emile C. Deal [Joel]. 1987. "Appendix 2: Lead Isotope Ratios." In Robert W. Bagley, *Shang Ritual Bronzes in the Arthur M. Sackler Collections*, 558–60. Ancient Chinese Bronzes in the Arthur M. Sackler Collections, vol. 1. Cambridge: Harvard University Press.

Barnes, I. Lynus, W. Thomas Chase, and Emile C. Joel. 1987. "Appendix 6: Lead Isotope Ratios." In Jessica. Rawson, *Western Zhou Ritual Bronzes from the Arthur M. Sackler Collections*, 168–71. Ancient Chinese Bronzes in the Arthur M. Sackler Collections, vols. 2A–B. Cambridge: Harvard University Press.

———. 1995. "Appendix 4: Lead Isotope Ratios." In Jenny So, *Eastern Zhou Ritual Bronzes from the Arthur M. Sackler Collections*, 489–91. Ancient Chinese Bronzes in the Arthur M. Sackler Collections, vol. 3. New York: Abrams.

Barnes, I. Lynus, W. Thomas Chase, Lore L. Holmes, Emile C. Joel, Pieter Meyers, and Edward V. Sayre. 1988. "The Technical Examination, Lead Isotope Determination, and Elemental Analysis of Some Shang and Zhou Dynasty Bronze Vessels." In *The Beginning of the Use of Metals and Alloys*, ed. Robert Maddin, 296–306. Cambridge: MIT Press.

Brill, Robert H., Yamasaki Kazuo, I. Lynus Barnes, K.J.R. Rosman, Migdalia Diaz. 1979. "Lead Isotopes in Some Japanese and Chinese Glasses." *Ars Orientalis* 11:87–109.

Brill, Robert H., William R. Shields, and J.M. Wampler. 1973. "New Directions in Lead Isotope Research." In *Application of Science in Examination of Works of Art*, ed. William J. Young, 73–83. Boston: Museum of Fine Arts.

Gale, N.H.. 1989. "Lead Isotope Analysis Applied to Provenance Studies: A Brief Review." In *Archaeometry: Proceedings of the 25th International Symposium*, ed. Y. Maniatis, 469–502. Amsterdam: Elsevier Science Publishers B.V.

Parkes, Penelope A. 1986. *Current Scientific Techniques in Archaeology*. New York: St. Martin's Press.

Sayre, Edward V., Paul Jett, and Emile C. Joel. 1992. "A Technical Examination of the Chinese Buddhist Bronzes in the Freer Gallery of Art, Part B: Stable Lead Isotope Analysis." In *Material Issues in Art and Archaeology III*, ed. Pamela B. Vandiver, et al., 225–37. Pittsburgh: Materials Research Society.

Yamasaki Kazuo, et al. 1979. "Lead Isotope Ratios in Some Japanese and Chinese Archaeological Bronzes." In *International Symposium on the Conservation and Restoration of Cultural Property: Cultural Property and Analytical Chemistry*, 222–34. Tokyo: Tokyo National Research Institute of Cultural Properties.

Index

Alexander, William Cleverly, 30 n.32

alloy, 98, 150, 152, 162, 173 n.2, 176 n.64, 179, 180, 181

American Art Galleries, 19, 20

Andersson, Johan Gunnar, 23

animal heads, 19, 47, 49, 83, 86 n.3, 95 n.2, 99, 113, 115, 115 n.1, 125
 on handles, 47, 49, 55, 67, 79, 83, 86 n.3, 87, 89, 90, 95, 103, 107, 109, 113, 121, 129, 131, 135, 139, 166, 170
 with antlers, 99, 107, 127, 168

animal motifs
 animals/zoomorphs, 39, 41, 71, 77, 78 n.1, 85, 87, 93, 97, 103, 129, 131, 135
 birds, 26, 43, 47, 48 n.3, 97, 98, 98 nn.8 & 10, 103, 107, 109, 113, 121, 121 n.3, 129, 137, 167
 bovine, 49, 67, 79, 83, 95, 135, 166, 170
 cicadas, 63, 99, 121
 feline, 115, 137, No. 29
 silkworms, 59, 61, 97
 snakes, 73, 73 n.2, 95, 97, 117, 159, No. 30
 See also dragons; animal heads

animal-shaped vessels, 27

Anyang, 15, 27, 28, 42 n.2, 49, 86 n.1, 95, 97, 98 n.5, 99, 101, 119, 119 n.3
 Dasikongcun, 51 n.2, 59, 59 nn.2 & 3, 67 n.3
 Gaolouzhuang, 59 n.3
 Hougang, 86 n.3
 Houjiazhuang, 42 n.3, 51 n.3, 83
 Qijiazhuang, 86 n.5
 Wuguancun, 45 nn.2 & 3
 Xiaotun, 51 n.2, 78 n.5, 83, 95 n.4; *see also* Fu Hao tomb
 Yinxu, 15, 83

applied products: 26, 27, 49, 148
 dirt/accretions, 158, 161, 163, 167
 fill, 154, 155, 163, 167, 172
 lacquer, 162
 overpaint, 154, 155, 158, 161, 162, 164, 165, 167, 172, Fig. 1.3
 waxy material, 155, 169, Figs. 1.3–4

archaisms, 13, 79, 90, 91 n.17, 131

Art Institute of Chicago, 25, 32 n.77, 125 n.2, 137

Arthur M. Sackler collections, 13, 28, 57 n.4, 67, 72, 72 n.4, 73 n.1, 81 n.1, 95 n.3, 121, 121 n.2, 123

Arthur M. Sackler Gallery. *See* Freer Gallery of Art and Arthur M. Sackler Gallery

Arthur M. Sackler Museum of Art and Archaeology, Beijing, 29

Arthur M. Sackler Museum, Harvard University, 57 n.2

Ashmolean Museum, Oxford, 48 n.4, 83 n.3

Asia House Gallery, 27

Asian Art Museum, San Francisco, 57 n.4, 101, 119, 133 n.2

azurite, 43, 47, 59, 63, 79, 87, 109, 131, 135, 151, 152, 154, 159, 162, 167, 170, 171, 172, 174 n.28, 179, 180, 181

Bagley, Robert W., 13, 28, 32 n.82, 89

Baicaopo, Lingtai Xian, Gansu province, 108 n.3

Baifu, Changping, 115, 115 n.1

Bang, Marquis of Jin, consort of, 86 n.1

Baoji altar sets, 26, 101
 Duanfang altar, 23, 26, 31 n.48
 Marquis of Qi bronzes, 26, 101
 "third" Baoji set, 101

Baoji Xian, Shaanxi, 101
 Daijiagou, Doujitai, Baoji Xian, 101

Beckmann, Dr. J. William, 12, 25

Beckmann, Leona J., 12

Beidongcun, Kezuo Xian, Liaoning province, 108 n.3

Beijing, 11, 18, 26, 78 n.3, 98 n.7

Beizhaocun, Qucunzhen, Quwo Xian, Shanxi, 133 n.3

Berlin State Museum, 17, 29 n.4

Bi Yuan, 105 n.2

Bigelow, William Sturgis, 20

Bing, Marcel, 21, 30 n.29

Bing, Siegfried (Samuel), 19, 21, 29 n.17, 139 n.1

Bixby, William K., 11

books on bronzes. *See* collection catalogues; reference works

British Museum, 23, 24, 31 n.66, 81 n.1, 93 n.1, 130, 130 n.7, 141 n.3

bronze, composition of. *See* alloy

bronze casting, 148–51, 179, Diagrams 1–2; *see also* chaplets; cores; model; mold

integral casting, 160, 162, 164, 166, 169, 180
precast elements, attachment, 131, 171,
precast elements, inserted in mold, 168
172, Figs. 10.3, 10.6, 10.7, 10.8

bronze disease, 151, 166

bronze porosity, 153, 155, 157, 158, 160, 164,
166, 170, 172

Buckingham, Kate Sturges, 25, 31 n.58

Buckingham, Lucy Maud, 25

Canton, 11, 18, 29 n.12

casting flash. *See* mold marks

Cernuschi, Henri, 17–18, 29 n.7

chamfrons, 115, 115 n.1, 125

Chang, Kwang-chih 32 n.82

chaplets, 19, 121, 148, 150, 153, 155, 156,
157, 158, 161, 162, 164, 165, 166, 167,
169, 170, 171, 172, 174 nn.16 & 18,
179, Diagram 2, Figs. 2.3–2.4, 4.6, 10.4
missing, 19, 161, 171, Figs. 4.8, 10.2–4

Chen Jieqi, 103, 105 n.5

Chen Mengjia, 25, 26–27, 32 nn.75 & 77, 89,
139 n.1

Cheung Kwong-yue, 32 n.77

Chiang Kai-shek, 24

Chu Deyi, 17

Chun, Prince, 19

City Art Museum of Saint Louis/The Saint
Louis Art Museum, 11, 12, 25, 26

Cleveland Museum of Art, 23, 133 n.2

Collection catalogues
American
Alfred F. Pillsbury collection, 25, 31 n.62
*Ancient Chinese Bronzes of the Shang
and Chou Dynasties*, 25
Art from Ritual, 28
Buckingham collection, 24
Freer collection, 24
Early Chinese Bronzes, 12, 25, 31 n.65
Ancient Chinese Bronzes (1987–95),
13, 28, 32 n.85
Chinese
Kankarō kikkinzu, 26
Shierjia jijin lu, 26
Taozhai jijin lu, 26, 31 n.47, 87, 101
Xiqing gujian, 18
European
Cull collection, 24
Chinesische Bronzen aus der Abteilung

für Ostasiatische Kunst, 17
Eumorfopoulos collection, 21
*Frühe chinesische Bronzen aus der
Sammlung Oskar Trautmann*, 24
*Sammlung Lochow: Chinesische
Bronzen*, 24
Japanese
Hakkakujō, 20
Sen'oku seishō, 20
*Shōsanrindō shoga bunbō
zuroku*, 17

cores, 19, 148, 153, 156, 174 n.12, 179, 180,
Diagram 1
construction of, 150
copper cores, 150, 166, 174 n.15, 176 n.62,
Fig. 7.4
core extensions, 150, 153, 154, 156, 159,
160, 161, 162, 163, 171, 176 n.67, 179,
Figs. 1.1–2, 2.2, 4.9, 4.10, 5.2, 9.1
cores within appendages, 135, 150, 160, 163,
165, 168, 170, Figs. 4.7, 4.9, 5.2, 9.6
definition of, 149
interleg core, 161, 162, 165, 179, 180,
Diagram 1
hollowed-out cores, 170, Fig. 9.7
metal in core material, 165, 166, Figs. 4.7,
7.5

corrosion processes, 151, 174 n.25
Type I (delta removal), 151, 172, 175 n.33
Type II (alpha removal), 151–52, 157, 164,
174 n.28, 181; *see also* water patina

corrosion products, 148, 151, 169, 181, Fig. 1.3
of copper, 151, 156, 162, 167, 169, 170, 171,
174 n.28, 179, 181; *see also* azurite;
cuprite; malachite
of lead, 151, 161, 169, 170, 174 n.28
of tin, 151, 162, 172, 174 n.28, 175 n.41,
181

covers, 47, 51, 55, 57, 57 n.3, 79, 85, 93, 99,
103, 11 n.5, 119, 129, 131, 133 n.1, 137

cross section analysis, 148

cruciform openings, 39, 43, 57 n.1, 154,
Figs. 1.1–2

cruciform recessions, 61, 156, Fig. 2.2

CT. *See* industrial computed tomography;
X-ray computed tomography

CT slice, 152, 153, 161, 166, 170, 172,
Figs. 4.7, 4.10, 7.4, 7.6, 9.4, 10.4, 10.7–8

Cull, Anders Eric Knös and James K., 24

cuprite, 39, 41, 43, 47, 49, 55, 59, 61, 63, 65,
67, 69, 71, 79, 83, 85, 87, 97, 103, 107, 109,
117, 119, 121, 131, 151, 152, 154, 156, 157,
158, 159, 162, 164, 165, 166, 167, 169, 171,
174 n.28, 179, 180, 181

cuprite marker layer, 174 n.28, 179

Dai, wife of the Marquis of, 141

danglu, 114–15, No. 29
shi danglu, from Baifu, 115 n.1

Dasikongcun, Anyang, 51 n.2, 59, 59 nn.2 &
3, 67 n.3

Davis, J. Lionberger, 12, 25, 26

Dayangzhou, Xin'gan, Jiangxi province, 28

Delbanco, Dawn Ho, 28

dendritic structures, 154, 157, 162, 164, 167,
172, 176 n.63, 179, 180, 181, Fig. 8.3

design transference
drawing or stamping, 159, 175 n.37, Fig. 4.2
model to mold, 148, 159, 162, 164, Fig. 4.2
mold to bronze, 154, 159, 162, 169,
173 n.3, Fig. 6.1

Di Xin, 15, 90, 90 n.3, 91 n.19

Di Yi, 15, 90, 90 n.3

digital radiograph (DR), 152, 153, 161, 166,
Figs. 4.9, 7.5, 9.6; *see also* industrial
computed tomography; X-ray computed
tomography

ding, 52–54, 62–63, 64–65, 68–69, 79,
179, Nos. 6, 10, 11, 13
in other collections, 54 n.1, 72 n.3, 101
in other references, 45 nn.2 & 3, 54 n.1,
65 n.1, 78 n.2, 135 n.3
trilobed *ding*, 62–63, 64–65, Nos. 10–11

Dongjiacun, Qishan Xian, Shaanxi, 127 n.1

dou, 133 n.4

dragons, 25, 43, 45, 47, 49, 53, 55, 57 n.2, 65,
73, 85, 97, 98 n.4, 99, 107, 108 nn.1 & 4,
109, 113, 121, 127, 131, 135, 135 n.3, 137,
139, 166, 167, Figs. 8.2–3

Duanfang, 11, 25, 31 n.47, 87, 101
widow of, 23

Duanfang altar. *See* Baoji altar sets, Duanfang
altar

Duret, Théodore, 17–18, 29 n.10

Ecke, Gustav, 24

Erligang (site), 15

Erlitou (site), 15

Eumorfopoulos, George, 20, 21, 22, 23

Exhibitions:
Asia House Gallery, New York, 1968, 12, 27

Exhibition of Chinese Art, Amsterdam,
  1925, 22
International Chinese Exhibition, London,
  1935–36, 18, 23, 24
Louisiana Purchase Exposition,
  Saint Louis, 1904, 11
Metropolitan Museum of Art, New York,
  1916, 21
Metropolitan Museum of Art, New York,
  1938, 23, 24
Metropolitan Musuem of Art, New York,
  1980, 28
Musée Cernuschi, Paris, 1937, 23
Musée de l'Orangerie, Paris, 1934, 22
Museum of Far Eastern Antiquities,
  Stockholm, 1933, 22
Oriental Ceramic Society, London, 1951, 18
Palais de l'Industrie, Paris, 1873–74, 18
Preussischen Akademie der Künste, Berlin,
  1929, 22, 30 n.35

fangding, 73–75, 76–78, 93, 95 n.4, 159–61,
  162–63, 164, 175 n.46, 176 n.67, 182,
  Nos. 15, 16, Figs. 5.1–2
  Geng Shi fangding, 78 n.5
  in other collections, 73, 78 nn.3 & 5
  in other references, 71, 78 n.3, 95 n.4

fanghu, 86 n.3, 133 n.3

fanglei, 12, 99–102, 153, 168–170, 182,
  No. 24, Figs. 9.1–7

fangyi, 84–86, 164, 182, No. 19, Fig. 6.1
  in other collections, 86 n.3, 101
  in other references, 86 nn.3 & 5

fangzun, 19

Fenollosa, Ernest F., 20

Ferguson, John C., 26, 32 n.83

Field Museum, 23

FitzWilliam Museum, 127 n.5

fluorescence under UV light, 148, 152, 153,
  155, 158, 161, 163, 166, 167, 169, 172

forgery, 26, 29 n.6, 31 n.75, 32 n.77, 97, 113

fou, 51 n.4

Freer, Charles Lang, 11, 19, 20, 21, 25, 26, 31
  n.60

Freer Gallery of Art, 12, 19, 21, 25, 32 n.77,
  113 n.1

Freer Gallery of Art and Arthur M. Sackler
  Gallery, 13, 89, 95 n.2, 97, 130 n.2, 139 n.1

FTIR analysis, 162, 164, 175 n.46

Fu Hao tomb, 78 n.3, 86 n.3, 102 n.1

Fufeng Xian, Shaanxi, 127

Gaojiabao, Jingyang Xian, Shaanxi,
  108 nn.2 & 4

Gaolouzhuang, Anyang, 59 n.3

GE Aircraft Engines Quality Technology
  Center, 152, 153

George V, King, 18

gong, 135

Gong, King, 127

Gong, Prince of Manchu, 20

Grousset, René, 23, 30 n.35

gu, 18, 38–39, 43–45, 47, 58–59, 60–61, 63,
  67 n.2, 86 n.6, 154–55, 156–57, 158,
  179, 181, 182, Nos. 1, 3, 8, 9,
  Figs. 1.1–4, 2.1–4
  in other collections, 18, 45 n.3
  in other references, 39, 59, 59 n.2, 98
  in pairs, 59, 59 n.3, 61, 61 n.2

Guangdong, 29 n.12

Guangxi, 29 n.12

Guangxu Emperor, 18–19

gui, 12, 25–26, 106–8, 109–11, 112–13,
  120–21, 131–33, 135, 171–72, 182,
  Nos. 26–28, 32, 37, Figs. 10.1–8
  in other collections, 25, 31 n.66, 111 n.5,
  121, 121 n.2
  in other references, 108 nn.2–4, 133 n.2,
  135 n.2
  in pairs, 108 n.2, 133 n.1
  Li *gui*, 109, 111 n.6
  Ya Chou *gui*, 111 n.5

Gustav Adolf, Prince, 22

Hakutsuru Bijutsukan, 20, 89, 105 n.7

Hamada Kōsaku, 20, 30, n.25

handle attachment, 171, 172, Figs. 10.3, 10.6–8

handles, 47
  animal as handle, 137
  bail, 55, 103
  C-shaped, 79, 95, 109, 119, 121
  closed ring, 83
  dragon and ring, 131
  inverted-U, 53, 63, 65, 69, 73, 77
  loop handle, 47, 49, 67
  lug and ring, 86 n.3, 99, 103, 129, 139
  lugs, 49, 86 n.3, 113
  strap handle, 41
  See also animal heads, on handles

Harada Yoshito, 30 n.25

Hart, Sir Robert, 11

Harvard University Art Museums, 83 n.3, 101

Hayashi Minao, 98 n.8, 121 n.2

Hayashi Tadamasa, 17

he, 18, 79, 135, 136–37, No. 39
  in other collections, 93 n.1, 137

Hearn, Maxwell K., 32 n.82

Hejiacun, Jishan Xian, Shaanxi, 108 n.2

Henan Provincial Museum, 78 n.3

Hoopes, Thomas Temple and Catherine
  Filsinger, 12

hot tearing, 180
  prevention of, 156, 157, 158

Hougang, Anyang, 86 n.3

Houjiazhuang, Anyang, 42 n.3, 51 n.3, 83

hu, 128–30, 131, 138–39, Nos. 36, 40
  in other collections, 32 n.77, 130,
  130 nn.2 & 7
  in other references, 45 n.2, 51 n.2, 130 n.7
  Ji Fu *hu*, 129, 130, 130 n.3
  San Nian Xing *hu*, 129, 130 n.5
  Zhao Meng *hu* (Cull *hu*), 24

Huang Jun, 24

Hubei province, 11

Huizong Emperor, 29 n.10

Hunan Provincial Museum, 57 n.3

Ichikawa Beian, 17

Idemitsu Bijutsukan, 105 n.7

industrial computed tomography (CT), 152,
  175 n.34; *see also* X- ray computed
  tomography

Industrial Institute, Beijing, 11

infrared spectroscopy, 148

inlay decoration
  silver, 27, 141
  lacquer, 162, 175 n.52

inscriptions, 27, 155, 158, 159, 163, 164, 166,
  167, 169, Figs. 1.3–4, 4.3, 5.2, 8.2–3;
  see also Inscription Index
  calligraphy and, 89
  clan signs, 39, 63 n.1, 65, 67, 71, 71 n.2,
  72 n.3, 73, 87, 89, 91 n.11, 95, 97, 101,
  127
  dedicatory, 42 n.1, 65, 71, 79, 77, 79, 87,
  89, 103, 113, 121, 127
  documentary, 87, 89, 109, 131
  incised into bronze, 65

inserted in mold, 166
oracular, 72 n.3, 89, 101
orientation on *ding*, 63, 63 n.5
stamped into bronze, 159

ivory, horn, bone carving, 41, 42 n.3, 45 n.2, 49, 51 n.3

Jannings, Werner, 24, 31 n.57

jia, 11, 25, 32 n.70, 63 n.3, 40–42, 67, 79, 87–91, 153, 165–66, 182, Nos. 2, 20, Figs. 7.1–6
in other collections, 42 n.1
in other references, 42 n.2

jiao, 46–48, 63, 95 n.2, No. 4
in other collections, 48 n.4, 89, 95 n.2

Jingjiecun, Lingshi, Shanxi, 59 n.3, 61 n.2

jue, 66–67, 94–95, Nos. 12, 22
in other collections, 67 n.3, 95 n.3
in other references, 42 n.5, 67 n.3, 95 nn.3 & 4
in pairs, 59

Kahn, Mrs. Otto, 23

Kanō Jihei, 20

Karlbeck lines, 165, 180

Karlbeck, Orvar, 23, 30 n.38, 180

Karlgren, Bernhard, 23, 25

Kelley, Charles Fabens, 25

Kidder, J. Edward, Jr., 12, 25, 26, 90 n.1, 98 n.8

Kümmel, Otto, 17

Kunkel, Mrs. Charles (Anna), 12

lead isotope analysis, 148, 152, 158

lead isotope ratios, 182–83

lei, 49–51, 99, 197, 182, No. 5
in other collections, 51, 130 n.2
in other references, 49, 51 nn.2 & 4, 108 n.3

leiwen, 14, 22, 49, 53, 55, 59, 61, 67, 69, 73, 77, 79, 87, 107, 119, 121, 123, 127, 129, 130, 130 nn.1 & 2, 137, 159, 167
irregular *leiwen*, 43, 45, 47, 59, 63, 65, 67; *see also* Loehr Style V[a]

li, 79

Liang Shangchun, 25

lids. *See* covers

lihe, 79–81, 111 n.5, No. 17
in other collections, 81 n.1

ling, 51 n.4

Lintong, Shaanxi, 111 n.6

Linyi, Shandong, 133 n.2

Liulihe, Beijing, 72 n.4, 91 n.11

Liushanzhai, Linqu, Shandong, 103, 105 n.2

"Liyu" bronzes, 22

Lodge, John Ellerton, 25

Loehr, Max, 12, 14, 23, 26, 39 n.1, 86 n.4, 98 nn.7 & 8

Loehr's Typology, 14, 27
Style I, 14, 42 n.2, 173 n.3
Style II, 14, 39, 42 n.2, 57 n.1, 79, 85, 173 n.3
Style III, 14, 39, 41, 42, 42 n.2, 43, 44, 77, 85, 95, 173 n.3
Style IV, 14, 42 nn.2 & 4, 43, 45 n.2, 49, 63, 173 n.3
Style V, 14, 43, 65. 173 n.3, 98 n.8
Style V[a]/*leiwen* relief, 59, 67, 67 nn.2 & 3

loess, 148, 150, 173 n.4, 179, 180

Loo , C.T., 12, 17, 25

Luoyang, Henan province, 131

Machida Hisanari, 29 n.4

marble carving, 119, 119 n.3

marble vessels, 83, 83 n.2, 109

maguan, 124–25, No. 34
in other collections. 125, 125 n.2

malachite, 39, 41, 43, 47, 49, 55, 63, 67, 69, 71, 77, 79, 83, 85, 87, 95, 97, 103, 107, 109, 117, 119, 121, 123, 125, 129, 131, 135, 137, 139, 151, 152, 154, 156, 157, 158, 159, 160, 162, 163, 164, 167, 169, 170, 171, 174 n.28, 176 n.69, 179, 180, 181

Mawangdui, Changsha, Hunan province, 141

Menzies, James M., 17, 101

Metropolitan Museum of Art, New York, 21, 23, 26, 28, 32 n.83, 101, 121 n.2

Meyer, Agnes Ernst and Eugene, Jr., 21

Minneapolis Institute of Arts, 23, 25

model, 148–49, 173 n.3, 180, Diagram 1
See also design transference, model to mold

model (tomography), 152, 153, 170, Fig. 9.7

mold, 148–50, 173 nn.3 & 4, 180, Diagrams 1–2
inscriptions inserted in, 166
mold inserts, 165, 171
See also design transference, mold to bronze; mold assembly

mold assembly, 150, 168, 180, Diagram 1
four-part divided at corners, 131, 159, 162, 164, 168, 171, Diagram 1
four-part divided at flanges, 71, 158
three-part, 53, 165
two-part, 39, 47, 59, 61, 67, 154, 156, 167, 171

mold flash. *See* mold marks

mold marks, 67, 71, 150, 174 n.21, 180
as lines/ridges, 150, 154, 156, 158, 159, 164, 165, 168, 171, 180, Figs. 1.1–2, 7.1, 9.1, 10.2
as misalignment of decoration, 150, 156, 164, 165, 168, 171, Figs. 2.1, 10.1
as steps, 150, 167, 171, Fig. 10.1
from interleg core pieces, 165
metal overflow, 168, 171, 172, Fig. 9.2
removal of, 150, 180
showing misalignment of core extensions, 154, Figs. 1.1–2
See also Karlbeck lines

Moore, Rufus E., 19, 25

Morse, Edward Sylvester, 20

Musée Cernuschi, Paris, 17, 23, 29 n.7, 121 n.2

Musée de l'Orangerie, 22

Musée Guimet, 121 n.2

Museum für Ostasiatische Kunst, Cologne, 31

Museum of Far Eastern Antiquities, Stockholm, 22, 111 n.5, 119 n.5, 125

Museum of Fine Arts, Boston, 32 n.77, 130 n.2

Nanluo, Lintong, Shaanxi, 127 n.1

nao, 29, 31 n.56

Naozhai, 24

National Palace Museum, Taiwan, 72 n.4

Nelson-Atkins Museum, Kansas City, 23

oracle bones, 72 n.3, 75 n.6, 89

original surface (of bronze), 150, 152, 166, 175 n.33, 179; *see also* cuprite marker layer

ornament (inlaid cylinder), 140–41, No. 41
in other collections, 141 n.3
in other references, 141 n.1

Palace Museum, Beijing, 26, 31 n.56, 73, 86
n.3, 97, 98 n.7, 133 n.2

Palmgren, Nils, 30 n.38

pan, 126–26, No. 35
in combination with *yi*, 135
in combination with *yi* and *he*, 127, 127 n.1
in other collections, 127 n.5
in other references, 91 n.11
Lu Fu Yu *pan*, 127

Pan Geng, 15, 45 n.3

Panlongcheng, 14, 39

patina, 26, 27, 180, Nos. 1–41; *see also* water
patina

patterns
diamond and boss, 121, 121 n.2
eye and diagonals, 93, 99, 127, 129
hooks, volutes, and curls, 39
of bosses, 77, 162
snake and boss, 73, 73 n.2
swirling hooks, 41
wave pattern, 129, 130, 130 nn.1 & 2, 131
See also leiwen

Peking. *See* Beijing

People's Republic of China, 12, 26

Pillsbury, Alfred F., 25

posts
square with boss caps, 41, 67, 95
round with conical caps, 87

pottery/ceramic vessels, 21, 51 n.2
and bronze design, 23, 45 n.2, 49, 51, 65,
79, 83, 90, 109, 130 n.2

pou, 73 n.1

Preussischen Akademie der Künste, Berlin, 22

Priest, Alan, 23

pseudomorphs, 67, 156, 160, 162, 164, 169,
171, 176 n.54, 181, Figs. 4.5, 5.1, 10.5

Pulun, Prince, 11

Puyi, Emperor, 11

Qi, daughter of the duke of, 131

Qi, Marquis of, 26, 85

Qianlong Emperor, 18, 29 n.11

Qijiacun, Fufeng Xian, Shaanxi, 130 n.3,
135 n.2

Qijiazhuang, Anyang, 86 n.5

Qishan Hejiacun Xihao, Shaanxi, 83, 83 n.3

Rawson, Jessica, 13, 28

Reference works:
Bogu tu, 18, 29 n.10
Fuzhai jijin lu, 103, 105 n.5
Haiwai jijin tulu, 21
Jiguzhai zhongding yiqi kuanshi fatie, 18,
29 n.12
Lidai zhongding yiqi kuanshi, 29 n.12
Nihon shūcho Shina kodō seika, 21
Ōbei shūcho Shina kodō seika, 20, 21
Shanzuo jinshi zhi, 103, 105 n.2
Xiqing gujian, 18
*Yu und Kuang: Zum Typologie der
chinesischen Bronzen*, 31 n.51

Renmin gongyuan, Zhengzhou, Henan
province, 45 n.2

repairs, 27, 47, 65, 98, 131, 153, 160, 167
adhesives, 172
fill, 154, 155, 163, 167, 169, 172
metal rods, 131, 161, 172, Figs. 4.7, 4.9–10
plug/patch, 161, 168, Figs. 4.8, 9.3–4
solder, 161, 162, 163, 172, Figs. 4.1, 4.6, 5.2

replacements, 27, 161, 171, Fig. 4.9

restoration, 39, 49, 131

Rong Geng, 21, 24

Rong Hou, 26

Roorda, T.B., 22

Royal Academy of Arts, London, 22

Royal Ontario Museum, 45 n.3, 95 n.3

Ruan Yuan, 29 n.12, 103, 105 nn.2 & 4

Sackler, Dr. Arthur M., 28; *see also* Arthur M.
Sackler collections

The Saint Louis Art Museum/City Art Museum
of Saint Louis, 11, 12, 25, 26

sales
American Art Galleries, New York, 1913, 20
American Art Galleries, New York, 1914, 20
Collection Hayashi, Paris, 1903, 17, 29 n.3
of bronze *you*, London, 1910, 22, 30 n.32
of Duanfang altar, 23
of Marcel Bing's collection, 21

Salles, Georges, 22

Sanxingdui, Guanghan, Sichuan province, 28

Seattle Art Museum, 57 n.3, 121 n.2

SEM analysis, 162, 175 nn.51 & 52

septum, 156, 157, 181, Figs. 2.3–4,

Shaanxi province, 19, 101, 107, 129
Baoji Xian, 101

Daijiagou, Doujitai, Baoji Xian, 101
Dongjiacun, Qishan Xian, 127 n.1
Doujitai, Baoji Xian, 101
Fufeng Xian, 127
Gaojiabao, Jingyang Xian, 108 nn.2 & 4
Hejiacun, Jishan Xian, 108 n.2
Lintong, 111 n.6
Nanluo, Lintong, 127 n.1
Shaochencun, Fufeng Xian, 135 n.2
Qishan Hejiacun Xihao, 83, 83 n.3
Qijiacun, Fufeng Xian, 130 n.3, 135 n.2
Xi'an, 127
Zhuangbai, Fufeng Xian, 130 n.5
Zichang, 39

Shandong province, 103, 105 nn. 2 & 5, 129,
130, 130 n.6

Shanghai Museum, 57 n.3

Shanxi province, 22, 61
Beizhaocun, Qucunzhen, Quwo Xian,
133 n.3
Jingjiecun, Lingshi, 59 n.3, 61 n.2
Taiyuan, 125

Shaochencun, Fufeng Xian, Shaanxi, 135 n.2

shengkeng bronzes, 27

Shengyu, 26

shoukeng bronzes, 26, 27

snake (bronze), 12, 116–17, No. 30
in other collections, 117

So, Jenny F., 13, 28, 32 n.82

spouts
channel, 67, 95, 135
bird-shaped, 137
tubular, 79

sprue, 161, 181, Figs. 4.6, 4.8

Stedelijk Museum, 22

stone ritual objects, 28

stress cracking, 164

stylistic chronologies of bronzes
Andersson, Karlgren, Karlbeck, 31 n.44
Chinese Organizing Committee, 23, 31 n.42
Loehr, 14, 27; *see also* Loehr's Typology
Karlbeck and Palmgren, 22, 30 n.38
Yetts, 30 n.42

Sumitomo Kichizaemon, Baron, 20, 21

surface alterations, 156, 27
black products, 69, 156, 159–60, 162,
164, 175 nn.40 & 50, Fig. 4.4
residual core material, 156, 171
textile remains, 156, 162, 164, 176 n.54,
Fig. 5.1

surface condition
    cleaning with acid, 158, 165, 169
    mechanical cleaning, 154, 164, 165, 167,
        176 n.69, Fig. 8.1
    pitted surface, Fig. 3.3
    scratches 155, 159, 162, 165, 166, 167,
        Figs. 1.3–4, 4.4, 7.3

Suzhou, 26, 32 n.77

Taixicun, Gaocheng, Hebei province, 39

Taiyuan, Shanxi, 125

taotie, 22, 26, 39, 41, 42 n.5, 45, 53, 55, 57,
    59, 61, 63, 63 n.5, 65, 67, 69, 71, 73, 77,
    78 n.1, 85, 86 n.7, 87, 95, 97, 98, 98 n.10,
    99, 103, 109, 113, 119, 125, 127, 135, 158,
    161

Thoms, Peter Perring, 29 nn.10 & 15

Thorp, Robert L., 32 n.82

tiaose qi, 82–83, No. 18
    in other collections, 83 n.3
    in other references, 83 nn.2 & 3

tomography. *See* industrial computed
    tomography; X-ray computed tomography

Trautmann, Dr. Oskar Paul, 24

Trübner, Dr. Jörg, 23, 31 n.51

Umehara Sueji, 20, 21, 32 n.70

van Heusden, Willem, 25

vessel walls
    buttresses, 156, 158, Figs. 2.3–4, 3.1
    stepped neck, 157, 158, Fig. 2.4
    thickening to strengthen vessel, 164, 168

Victoria, Queen, 18

Victoria and Albert Museum, 19, 23, 105 n.7

von Ketteler, Baron Klemens, 18

von Lochow, Hans Jürgen, 24, 31

Wang Chen, 26

Wannieck, Leon, 22

water patina, 152, 154, 156, 159, 162, 164,
    165, 167, 169, 171, 175 nn.31 & 33, 179,
    180, 181

Wen, Marquis of Jin, 86 n.1

Wen (Wu) Ding, 15, 86 nn.3 & 5

Wenxian, Henan province, 71

Wilhelm II, Kaiser, 18–19, 21

Wing Cheong and Company, 11

Worch, Edgar, 24

Wu Shifen, 105 n.5

Wu Wang, 109

Wu Yi, 15, 86 n.5

Wuguancun, Anyang, 45 nn.2 & 3

X-radiography, 131, 148, 152, 153, 155, 157,
    158, 160, 162, 163, 164, 166, 167, 170,
    172, 174 n.16, Figs. 4.6, 5.2, 9.5–6

X-ray computed tomography, 148, 152, 166,
    170, 172; *see also* industrial computed
    tomography
    images produced from, 152–53; *see also*
        CT slice; digital radiograph; model
        (tomography)

X-ray fluorescence, 148, 152, 164

X-ray powder diffraction, 148, 152, 164, 167

Xi'an, Shaanxi, 127

Xiao Xin and Xiao Yi, 45 n.3

Xiaoliuzhuang, Guicheng, Huang Xian,
    Shandong, 130 n.1

Xiaotun, Anyang, 51 n.2, 78 n.5, 83, 95 n.4

Xicun, Luoyang, Henan province, 117

Xu Jinxiung, 89, 90, 91 n.19

Xuan Wang, 133 n.3

Xue Shanggong, 29 n.12

Xun Xian, 125

Yamanaka Sadajirō, 20, 21, 26, 30 n.21

Yang Xiung, 101

yangzhuang bronzes, 27

Yetts, W. Perceval, 22, 24, 30 n.42

yi, 134–35, No. 38
    in combination with *pan*, 135
    in combination with *pan* and *he*, 127,
        127 n.1
    in other references, 127 n.1, 135 n.2

Yi Xian, Hebei province, 26

Yinxu, 15, 83
    kings of, 15

Yinxu periods, 15, 61
    period I, 42, 45 n.3
    period II, 42, 78 n.3, 95 n.4
    period III, 59 n.3, 86 n.5
    period IV, 86 n.3, 95 n.4

you, 22, 55–57, 103–5, 107, Nos. 7, 25
    in other collections, 23, 89, 57 nn.1–4,
        105 n.7
    in other references, 30 n.32, 57 n.5,
        108 nn.2 & 4, 130 n.1
    types of *you*, 55, 57 n.1

yu, 19, 130 n.2
    Qi Hou *yu*, 131, 133

Zhangzhai Nanjie, Zhengzhou,
    Henan province, 78 n.3

zhi, 57 n.1, 73, 75 n.4, 92–93, 96–98, 118–19,
    122–23, 167, Nos. 21, 23, 31, 33,
    Figs. 8.1–3
    in other collections, 26, 32 n.75, 98 n.7,
        119, 119 n.5, 123
    in other references, 26, 57 n.1, 98 n.7
    in pairs, 119, 119 n.1
    Zi fu zhi, 97

Zhongcuigong, Palace Museum, Beijing, 24

Zhou Meiku, 32 n.77

Zhuangbai, Fufeng Xian, Shaanxi, 130 n.5

Zhuwajie, Peng Xian, Sichuan province,
    108 n.3

Zichang, Shaanxi, 39

zoomorphs. *See* animal motifs,
    animals/zoomorphs

zun, 51, 70–72, 107, 157, 158, 182, No. 14,
    Figs. 3.1–3
    in other collections, 23, 72 n.4
    in other references, 72 n.4, 108 n.2, 130 n.1
    in pairs, 119, 119 n.1

Inscription Index

Inscriptions from catalogued bronzes are listed in order by catalogue number. Pages where rubbings appear are in bold. Citations for individual graphs are given in alphabetical order, followed by undeciphered graphs (*X*).

X (No. 1; 29:1951), **39**

Jiu Yi (No. 2; 224:1950), **41**, 42 n.1

Niao (No. 3; 28:1951), **43**, 47

Tian Ding (No. 4; 220:1950), **47**, 48 n.1

Yin ban (No. 6; 31:1951), **53**, 54 nn.2 & 3

X (No. 9; 463:1956), **61**

X Gui (No. 10; 22:1049), **63**, 63 n.1

Shi Fu Xin (No. 11; 288:1955), **65**

Ya Xin (No. 12; 344:1955), **67**, 67 n.1

Zi Zu Xin Bu (No. 14; 125:1951), **71**, 71 n.2, 72 nn.3 & 4

X (No. 15; 108:1954), **73**, 73 n.3, 74 nn.4–6, 159, Fig. 4.3

Xiaozi zuo Fu Ji (No. 16; 222:1950), **77**, 78 n.2

Yi Fu X (No. 17; 287:1955), **79**

X Fu Yi (No. 19; 127:1951), **85**, 86 n.2

Guisi Wang xi Xiaochen Yi bei shipeng yongzuo Mu Gui zunyi wei Wang liusi yongri zai siyue Ya Yi (No. 20; 221:1950), 26, 31 n.69, 32 n.70, **87**, 89, 90–91 nn.1–16, 166, Fig. 7.3

X (No. 21; 30:1985), **93**, 93 nn.1–3; *see also X* (No. 15)

Ge Xin Zu (No. 22; 27:1985), **95**

Long (No. 23; 215:1950), **97**, 98 n.4, 167, Figs. 8.2–3

X (No. 24; 2:1941), **99**, 101, 102 nn.3–8, 169

Shi Po Zhi zuo Fu Gui yi (No. 25; 225:1950), **103**, 105 nn.2, 5 & 6

Zuo bao zun yi (No. 28; 540:1956), **113**, 113 n.1

Yuan zuo lüyi (No. 31; 27:1951), **119**, 119 n.2

Zuo Fu Yi Zi (No. 32; 9:1970), **121**

X (No. 33; 230:1954), **123**, 123 n.1

Zuo Zhongfei zun Ge (No. 35; 126:1951), **127**, 127 nn.3 & 4

X Zi Shu Gou zi zuo yi yi wannian yong zhi (No. 38; 213:1950), **135**

Graphs in Catalogued Items

ban, 53, 54 nn.2 & 3

bao, 113

bei, 87, 89

Bu, 71, 71 n.1, 72 nn.3 & 4

Ding, 47, 48 n.1

Fu, 32 n.75, 65, 77, 78 n.2, 79, 85, 98 n.4, 103, 121

Ge, 95, 127, 127 n.4

Gou, 135

Gui. 26. 87, 89, 90 n.7, 103

Guisi, 26, 87, 89, 90 nn.2 & 8

Ji, 77, 78 n.2

Jiu, 41, 42 n.1

liusi, 26, 87, 89, 90 n.8

Long, 97, 98 n.4, 167, Figs. 8.2–3

lüyi, 119, 119 n.2

Mu, 26, 87, 89, 90 n.7

Niao, 43

Po, 103

Shi, 65, 103

shipeng, 87, 89, 90 n.6

Shu, 135

siyue, 87, 89

Tian, 47, 48 n.1

Wang, 87, 89, 90 n.3

wannian, 135

wei, 87, 89

xi, 87, 89

Xiaochen, 26, 87, 89, 90 nn.4–5

Xiaozi, 77, 78 n.2

Xin, 65, 67, 71, 95

Ya (*yaxing*), 26, 67, 67 n.1, 87, 89, 90 n.10

Yi, 26, 41, 79, 85, 87, 89, 90 nn.5 & 7, 91 n.11, 121

yi, 103, 105 n.6, 113, 119 n.2, 135,

Yin, 53, 54 n.2

yong, 135

yongri, 26, 87, 89, 90 n.9

yongzuo, 87 , 89

Yuan, 119

zai, 87, 89

Zhi, 103

zhi, 135

Zhongfei, 127, 127 n.3

Zi, 71, 71 n.1, 121, 135

zi, 32 n.75, 135

Zu, 95

zun, 113, 127

zunyi, 87, 89

zuo, 77, 78 n.2, 103, 113, 119, 121, 127, 135

X (No.1), 39, 155, Figs. 1.3–4

X (No. 9), 61

X (No. 10) 63, 63 n.1

X (No. 15), 73, 73 n.3, 74 nn.4–6, 159, Fig. 4.3; *see also X* (No. 21)

X (No. 17), 79

X (No. 19), 85, 86 n.2

X (No. 21), 93, 93 nn. 1–2; *see also X* (No. 15)

X (No. 24), 99, 101, 102 nn.3–8

X (No. 33), 123, 123 n.1